I0816221

CHEMISTRY

100 IDEAS IN 100 WORDS

A whistle-stop tour of key concepts

CHEMISTRY 100 IDEAS IN 100 WORDS

A whistle-stop tour of key concepts

Adrian Dingle

Contents

Introduction

The study of the matter that makes up the world around us is as old as time itself. Chemical elements such as gold, silver, copper, tin, lead, iron, mercury, zinc and arsenic have all been known since prehistoric times and have helped to form civilization itself. Today, our lives are made safer, more comfortable, more exciting and more fulfilled by advancements in chemistry – from the medications we take to heating our homes, and from the phones we use to the food in our cupboards.

Chemistry has its roots among the great Greek philosophers such as Empedocles and Democritus. They sought to explain the world as they observed it, but without the benefit of a systematic or quantifiable base provided by scientific experimentation. Their ideas included the four classical elements and atomism, which show early explorations of the ideas that would later become elements and atoms.

The alchemists – in pursuit of creating gold and the elixir of life – combined various disciplines, such as philosophy, astronomy and religion, with a degree of experimental technique, but alchemy still lacked the strict discipline of modern science. Gradually, more formal analytical techniques entered the work. It was prodding at the edges of the chemistry we know today.

Alchemy gained momentum through the centuries until the middle of the 17th century and Robert Boyle. His seminal work *The Sceptical Chymist* is considered by many to mark the dawn of chemistry as a fundamental science in its own right due to his rejection of the four classical elements and his pioneering definition of elements and compounds. When Antoine Lavoisier and others developed the quantitative aspects of chemistry 100 years later, chemists could finally start to measure, via experiments, what they thought they knew. At that point, chemical theories could be proven with experimental, measurable and reproducible results.

Women are noticeably underrepresented in this book. As an overview of the most foundational discoveries in chemistry, it is evident that for most of the discipline's history, women's lack of access to formal education leaves the history of chemistry exceptionally male-focused. However, the study of chemistry is now much more balanced and hopefully the future of chemical innovation will continue to move in that direction.

Scientists have now identified and named the 118 chemical elements that make up everything in the known universe, and period 7 (the seventh row of the periodic table) is now complete. Does this mark the end of chemical discovery? Of course not, and the future, in terms of new materials, biochemical advances, sustainable and renewable energy and nanotechnology, is only just beginning. Wherever chemical discovery and development takes us next will be limited only by human imagination and endeavour – and neither is going to be curtailed anytime soon.

Atomic Theories

At its very core, chemistry relies upon an understanding of the basic building block of all matter: the atom. The history of the development of that understanding is a fascinating example of how scientific ideas evolve over time, with new experimental data leading to the refinement of earlier theories and a deeper comprehension of a world that is invisible in our daily lives.

The sheer density of discovery and advancement that occurred in the approximate 25-year period that started in the last few years of the 19th century is truly mind-boggling. As you'll see, the astonishing breakthroughs of the scientists of that time were in turn only possible because of the ideas that had been formulated in the centuries prior – that's how science works: it's a continuum of evidence, applied to older ideas, in order to create new, further understanding.

1

In 100 words

Greek philosophers proposed ideas to explain how the natural world worked. Empedocles advanced the theory of **four elements** – air, fire, water and earth – being the constituents of all matter. He suggested that any given material contained differing ratios of these four classical **elements**, making them all unique. Democritus attempted to address the fact that when a substance was cut open, it only resembled smaller pieces of the original, and never any of the four elements. He proposed the idea of atomism from *atomos*, meaning "indivisible". Democritus said that the *atomos* of any given substance were unique to it.

Earliest ideas

Chemistry is a science. A science is based upon observable natural phenomena that lead to hypotheses that can, in turn, be tested via experiments. The earliest thinkers regarding the natural world were philosophers, not scientists, and their ideas were just that – ideas. Moreover, they were ideas that were never tested by experiment. As such, belief systems rather than experimentally verified theories and laws became the accepted "understanding" of the day. Democritus's ideas about **atoms** were revolutionary in as much as they formed the basis for modern atomic theory. They were ultimately cemented by John Dalton and his atomic theory more than 2,000 years later.

WHY IT MATTERS
The desire of humans to understand the natural world has been present since the beginning of time

KEY THINKERS
Empedocles
(c. 490–c. 430 BCE)
Democritus
(c. 460–c. 370 BCE)

WHAT COMES NEXT
The basis of modern atomic theory has its roots in ancient Greek philosophy

SEE ALSO
Alchemy p.10
John Dalton's atomic theory p.18

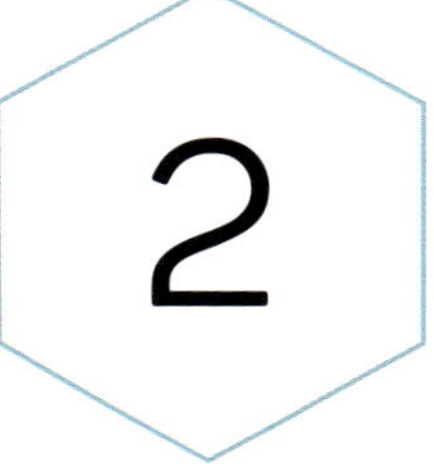

2 Alchemy

Abū Bakr Muhammad ibn Zakariyyā al-Rāzī is an interesting figure in **alchemy.** Many of his writings suggest a move away from explanations based on religion and sorcery, towards a more analytical and rational approach to finding answers. This is emphasized by the fact that his book *Kitāb al-Sirr al-asrār* (*The Book of the Secret of Secrets*) gives an extensive list of the equipment and apparatus that he used in his laboratory. Many of these items are recognizable as modern laboratory instruments, such as tongs, crucibles and flasks.

WHY IT MATTERS
Alchemy was the branch of natural philosophy that is considered a precursor to modern chemistry

KEY THINKER
Abū Bakr Muhammad ibn Zakariyyā al-Rāzī (865–925)

WHAT COMES NEXT
Philosophical ponderings evolved into a real science with the advent of quantifiable and repeatable experiments

SEE ALSO
The conservation of mass p.16

In 100 words

Alchemy was practised in various forms across the ancient world. A protoscience, it was the forerunner to the modern science of chemistry. It was mostly performed without formal, logical experimentation, whereas modern chemistry is the fully fledged experimental, quantifiable science that naturally followed from early philosophical ideas. Most famously, the purpose of alchemy was chiefly to convert base metals like lead into noble metals, such as gold, and to find the "elixir of life", which would enable eternal life. These goals were thought achievable if alchemists could only find the mythical substance called the philosopher's stone.

When the Anglo-Irish natural philosopher Robert Boyle published *The Sceptical Chymist* in 1661, he paved the way for much of modern chemistry. His theories would still need the work of Lavoisier, Priestley, Proust, Dalton and others in order to mature into what we recognize as chemistry today, but there are enough concepts and hypotheses in his work for many to consider it the seminal moment in the creation of the science. In his book, Boyle dismisses the Greek theory of the **four classical elements** and also redefines the terms ***element*** and ***compound*** in a manner that are recognizable today.

3

Robert Boyle

Boyle is perhaps best known for Boyle's law that relates the pressure and volume of a gas, and which is, strictly speaking, a notion rooted in physics. He was a prolific experimenter in a number of fields. Modern historians argue that *The Sceptical Chymist* wasn't really a rejection of **alchemy** and that, at the time, there really was no conflict between natural philosophy and what would become the real science of chemistry. They may well be correct, but Boyle's work was sufficiently insightful in terms of moving away from the classical four elements, towards a more modern definition, and so it has come to represent a split between alchemy and chemistry.

WHY IT MATTERS
Robert Boyle's book *The Sceptical Chymist*, considered by many to be the foundation of modern chemistry, contained fundamental ideas that are still recognizable today

KEY THINKER
Robert Boyle (1627–1691)

WHAT COMES NEXT
The move to serious experimental and quantitative chemistry by Antoine Lavoisier

SEE ALSO
Earliest ideas p.9
The conservation of mass p.16

4

Cathode rays

WHY IT MATTERS
The electron would never have been discovered without the initial work by Faraday on cathode rays

KEY THINKERS
Michael Faraday (1791–1867)
Sir William Crookes (1832–1919)

WHAT COMES NEXT
The establishment of "something" in cathode rays made the discovery of the electron possible

SEE ALSO
The plum pudding model p.22

Further improvements in the **vacuum** made by German glassblower Heinrich Geissler, physicist Julius Plücker and British chemist Sir William Crookes led to further experimentation. Crookes's tubes had better vacuums, and, in his tubes, Faraday dark space was observed further away from the cathode. German physicist Johann Hittorf suggested that observed shadows cast by the arc were due to something travelling from the cathode to the anode, and German physicist Eugen Goldstein suggested the name cathode rays. The rays became the basis for J J Thomson's research and, ultimately, the discovery of the **electron**.

"It is the chemist who must come to the rescue of the threatened communities. It is through the laboratory that starvation may ultimately be turned to plenty."
Sir William Crookes, chemist (1898)

In 100 words

One of the earliest experiments by the English scientist Michael Faraday laid much of the groundwork for the discovery of the first subatomic particle, the **electron** by J J Thomson in 1897 (see p.23). In 1838, Faraday passed a current through a tube that contained rarefied air – air that had most of the gas **atoms** removed, but where the **vacuum** was incomplete. He observed an arc of electricity starting at the negative plate (cathode) and passing to the positive plate (anode). There was a gap in the arc, which became known as Faraday dark space.

CROOKES VACUUM TUBE
Cathode rays, later identified as streams of electrons, were first detected in the partially evacuated Crookes tube.

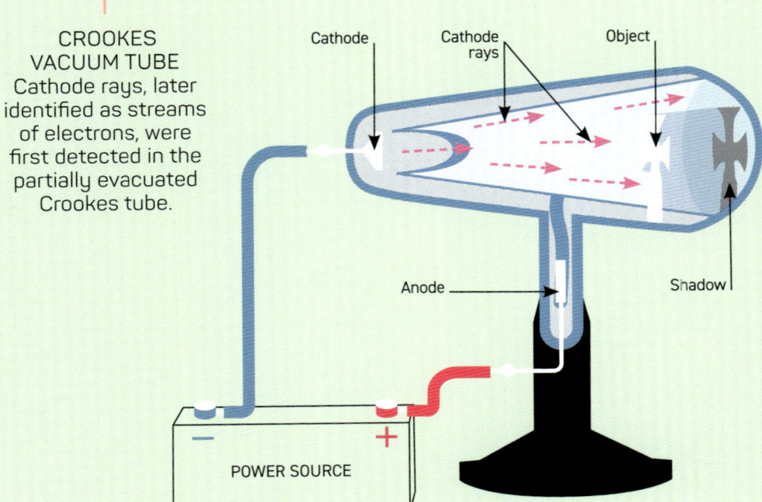

Phlogiston

WHY IT MATTERS
The phlogiston postulate was a forerunner to the discovery of oxygen, the development of quantitative chemistry and the law of conservation of mass

KEY THINKERS
Johann Becher (1635–1682)
Georg Ernst Stahl (1659–1734)
Joseph Priestley (1733–1804)
Antoine Lavoisier (1743–1794)

WHAT COMES NEXT
Alchemy evolves into chemistry – a quantitative science based on observation, measurement and experimentation

SEE ALSO
Alchemy p.10
The of conservation of mass p.16
Combustion reactions p.92

In 1774, via a series of experiments involving the suffocation of mice and the decomposition of mercuric oxide, English chemist Joseph Priestley described any atmosphere that could support combustion as being "dephlogisticated air". He based the name on the fact that when such a gas was present, it could continue to accept phlogiston from the substance being burned, and therefore it could continue to support combustion. Priestley hypothesized that once the dephlogisticated air had become saturated with phlogiston, the combustion process could not continue.

In the 1770s, quantitative experiments by the French chemist Antoine Lavoisier demonstrated that metals did not lose mass during combustion as the phlogiston theory predicted, but rather they actually gained mass instead. This dealt the fatal blow to the phlogiston theory, debunking the idea that phlogiston was lost during combustion, and instead suggested that a substance was gained by the combustible material. Lavoisier later identified Priestley's "dephlogisticated air" as an **element** and named it oxygen in 1777.

The quantitative aspect of much of Lavoisier's work helped to complete the transition from the myth and legend of **alchemy** to the hard science of chemistry, based on experimentation and measurement.

In 100 words

The phlogiston hypothesis was an early idea in the development of chemistry which was proposed in order to explain observations related to the combustion (burning) of substances. The belief postulated that all combustible materials contained a substance called phlogiston, which was released on burning and was absorbed by the surrounding atmosphere. First proposed by Johann Becher in 1669 via what he called *terra pinguis*, the notion was advanced in 1703 by Georg Ernst Stahl, who coined the term *phlogiston*. The theory was widely accepted, and it persisted until the quantitative work by Antoine Lavoisier 70 years later and the discovery of oxygen.

LAVOISIER'S APPARATUS
Lavoisier's experiments showed that metals gained mass during combustion, debunking the phlogiston theory.

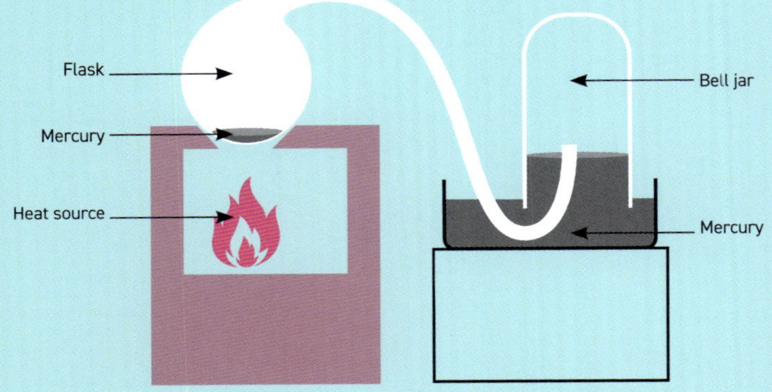

6

The conservation of mass

WHY IT MATTERS
Demonstrating that the mass of substances increased when they were burned finally brought the phlogiston theory to an end

KEY THINKERS
Antoine Lavoisier (1743–1794)
John Dalton (1766–1844)

WHAT COMES NEXT
Dalton's advanced atomic theory a few decades later

SEE ALSO
John Dalton's atomic theory p.18
The elements p.31

Lavoisier's success was based in the quantitative measurement of the mass of substances when they were burned. He observed that both phosphorus and sulphur each gained mass after combustion. This suggested to him that there was a combination of substances during such a reaction, and not phlogiston (see p.14) being used up, as had been the prevailing theory. In addition to all of his other achievements, Lavoisier also proposed a system of chemical nomenclature. Unfortunately, all of his genius could not save him from the French Revolution, where his role in tax collection for the monarchy led to him succumbing to the blade of the guillotine in Paris in 1794.

In 100 words

In 1789, French chemist and aristocrat Antonie Lavoisier put forth a number of important ideas in his book *Traité élémentaire de chimie (Elements of Chemistry)* – often considered the first chemistry textbook. In it, he suggested the first formal definition of the term ***element***, listed 33 such elements (only 23 of which fit our current understanding of the word), proposed the idea of a **chemical equation**, and stated the law of conservation of mass as: "Nothing is lost, nothing is created, everything is transformed." Today, the law states that matter cannot be created or destroyed, rather only converted from one form to another.

When James Chadwick discovered the **neutron** in 1932, he completed the triumvirate of subatomic particles: **electron**, **proton** and **neutron**. Chadwick conducted experiments that involved bombarding beryllium with alpha particles. His work was inspired by Walther Bothe, Herbert Becker and Frédéric and Irène Joliot-Curie, who had observed a form of unidentified, highly penetrating radiation during their similar experiments. Chadwick didn't believe that the radiation observed in those experiments was the gamma radiation that they had hypothesized, and his experiments concluded that a neutral particle, with a mass similar to that of a proton, was actually the source of the mysterious radiation.

7

Chadwick and the neutron

In 1912, J J Thomson observed some peculiar behaviour of neon, detecting two different masses for neon **ions**. At this time, the assumption was that **atoms** were made up of protons and electrons, but in 1920 Rutherford had hypothesized about the existence of some kind of neutral particle. The discovery of the neutron, and how the nuclei of atoms could contain varying numbers of them, explained many of those earlier observations. The discovery of the neutral particle also made sense of Francis William Aston's "whole number rule", which showed that most **atomic masses** were close to whole numbers. Chadwick was awarded for his discovery with the Nobel Prize in Physics in 1935.

WHY IT MATTERS
The neutron's discovery had consequences for multiple concepts in chemistry, including isotopes, mass spectrometry and radioactivity

KEY THINKER
James Chadwick (1891–1974)

WHAT COMES NEXT
In the 1960s, quarks (see p.168) were proposed as the elementary particles that make up neutrons

SEE ALSO
The plum pudding model p.22
The nuclear model p.24
Quarks and the Standard Model p.168

8

John Dalton's atomic theory

John Dalton's work in England in the very early part of the 19th century pulled together several earlier and slightly disparate ideas of many key thinkers. In the process, he formed a bridge between some of the earliest ideas about atomic theory and newer ideas, creating what became the basis of our modern understanding. A school teacher by trade, Dalton was a keen meteorologist. His observations of the rain, clouds and fog in the typical damp, maritime climate of England's Lake District, where he was born, grew up and worked, allowed him to formulate theories about air, the water vapour in it and how the gases mixed together and could change phase. It was these early observations that led him to his key work on atomic theory.

WHY IT MATTERS
Dalton's ideas cemented atomism and formed the basis of all of the modern atomic theory

KEY THINKER
John Dalton (1766–1844)

WHAT COMES NEXT
Dalton's theory was adjusted to account for isotopes (see p.20)

SEE ALSO
The conservation of mass p.16

In 100 words

Dalton's ideas were collected in *A New System of Chemical Philosophy* (published between 1808 and 1827). With its origins in Greek atomism and Joseph Proust, Antoine Lavoisier and Joseph Priestley's work, Dalton's theory states that all matter is composed of indivisible particles called *atoms*. He also explained that all the **atoms** of any given **element** are identical; all elements have their own unique atoms; atoms of different elements combine to form **compounds**; a compound always has the same number and type of atoms and atoms cannot be created or destroyed in chemical reactions, rather they are rearranged to form new compounds.

In 100 words

In 1938, when Otto Hahn and Fritz Strassmann found that barium was produced by the bombardment of uranium with **neutrons**, little could they have imagined the profound effect on humanity that their discovery would have. The process they had unearthed was **nuclear fission** – splitting a relatively large atom's **nucleus** into smaller nuclei, along with the release of an absolutely colossal amount of energy. German physicist Lise Meitner and her nephew Otto Frisch made sense of the observations, and it very quickly became apparent that fission would be central to the development of both nuclear power and nuclear weapons.

9

WHY IT MATTERS
The impact of nuclear power and nuclear war on the world is hard to overstate

KEY THINKERS
Otto Hahn (1879–1968)
Lise Meitner (1878–1968)
Otto Frisch (1904–1979)

WHAT COMES NEXT
The eventual creation of nuclear bombs

SEE ALSO
The naming of the elements p.32
Radioactivity p.130

Nuclear fission

A few years before Hahn's discovery, a similar experiment was conducted by Enrico Fermi, who also built the first functioning nuclear reactor. It was found that splitting a uranium **atom** with neutrons produced not only barium but also additional neutrons. These neutrons could be harvested to split more uranium and so on, thus creating the first nuclear chain reaction.

Otto Hahn was awarded the Nobel Prize in Chemistry in 1944, but his award has been shrouded in controversy because Lise Meitner's contributions were not recognized. However, in 1997, Meitner was honoured when **element** 109 (meitnerium) was named after her.

10

Isotopes

WHY IT MATTERS
The identification of isotopes is important in many aspects of chemistry, including nuclear medicine

KEY THINKERS
Frederick Soddy (1877–1956)
James Chadwick (1891–1974)
Francis William Aston (1877–1945)

WHAT COMES NEXT
The identification of new isotopes of already identified elements and the discovery of new elements (see p.34)

SEE ALSO
Chadwick and the neutron p.17

British chemist Frederick Soddy noticed that some **atoms** of radioactive **elements** exhibited identical chemical properties, but that they were different in terms of their **radioactivity**. By 1910, Soddy had established that mesothorium (now known as radium-228) was chemically identical to radium but that it exhibited different **atomic weights**. The word **isotope** combines the Greek words *iso* (meaning "the same") and *topos* (meaning "the place") to describe atoms that were different in some way but would have to occupy the same position in the **periodic table**. The existence of isotopes meant that Dalton's atomic theory (see p.18), which assumed all atoms of an element were identical, needed to be amended. Soddy was awarded the Nobel Prize in Chemistry in 1921 for his work. In 1919, British physicist Francis William Aston developed a machine called a mass spectrograph, which was the forerunner of the modern mass spectrometer. Aston used both electrical and magnetic fields to focus the **ion** beam. With it, he was able to confirm the existence of isotopes in nonradioactive elements, such as neon. Aston won the 1922 Nobel Prize in Chemistry, and over the course of his career he identified 212 isotopes.

In 100 words

Mass spectrometry is a technique used for the analysis of the isotopic composition of **elements** and for determining the structure of organic **compounds**. Isotopes are **atoms** of the same element (the same number of **protons**) but with differing numbers of **neutrons**. This gives them varying **atomic masses**. A mass spectrometer works by ionizing the source material and then separating the various chemical species based upon their mass-to-charge ratio. The resulting output, known as a mass spectrum, shows the relative abundance of the **ions** and their mass-to-charge ratios as peaks, with each ion showing up as a distinct line on the spectrum.

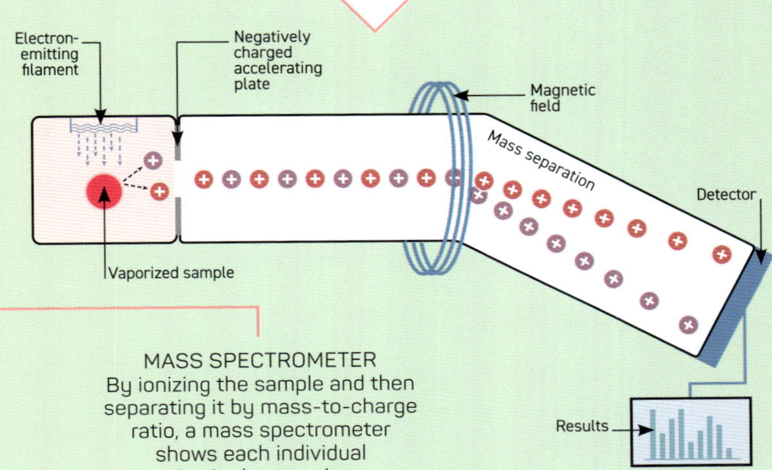

MASS SPECTROMETER
By ionizing the sample and then separating it by mass-to-charge ratio, a mass spectrometer shows each individual ion in the sample.

The plum pudding model

WHY IT MATTERS
The discovery of the first subatomic particle

KEY THINKER
Joseph John Thomson (1856–1940)

WHAT COMES NEXT
With the idea of subatomic particles in place, Rutherford went on to discover the proton and advance atomic theory

SEE ALSO
Chadwick and the neutron p.17
The nuclear model p.24

The discovery of the **electron** allowed J J Thomson to hypothesize more about the nature of the **atom.** Now knowing that there were subatomic particles and that there had to be some kind of positive charge to cancel out the negative electrons, he proposed the **plum pudding model** in 1904. The **model** resembled the dessert traditionally eaten in Britain on Christmas Day. In a plum pudding, a dough was interspersed with lots of raisins. Thomson said that the negatively charged electrons (the raisins) were randomly spread out among what he called "a sphere of uniform positive electrification" (the dough). Thomson won the Nobel Prize in Physics in 1906 for his investigations on the conduction of electricity by gases. The plum pudding atomic model was proven false by Rutherford's discovery of the atom's **nucleus** in 1911.

"Atoms are not indivisible, for negatively electrified particles can be torn from them by the action of electrical forces."

J J Thomson, physicist (1936)

In 100 words

Like Michael Faraday, Heinrich Geissler and William Crookes before him, J J Thomson studied cathode rays. Thomson found that the rays were deflected by magnetic fields and were attracted towards a positive electrical plate, meaning they were negatively charged. When the mass-to-charge ratio of the particles was found to be more than 1,000 times smaller than that of the smallest **atom** (hydrogen), Thomson knew he had found a subatomic particle. In 1897, Thomson announced his results and called these new negative particles *corpuscles*. Later the particles were renamed **electrons** – a word proposed by George Stoney in 1891 to describe the fundamental unit of electrical charge.

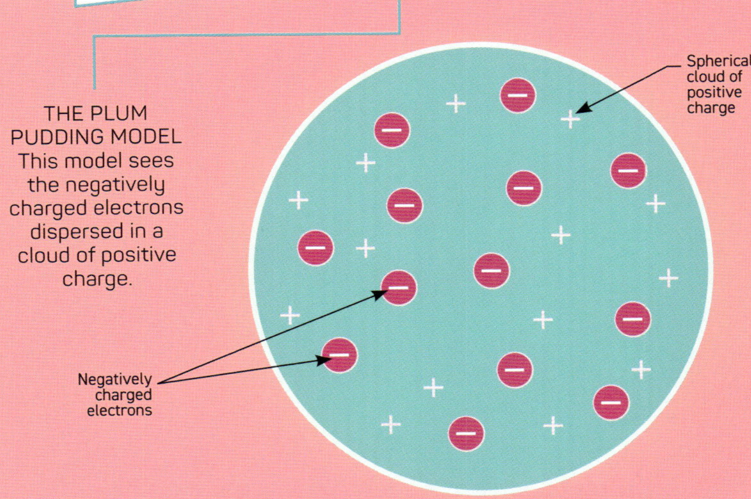

THE PLUM PUDDING MODEL
This model sees the negatively charged electrons dispersed in a cloud of positive charge.

12

The nuclear model

WHY IT MATTERS
The development of the nuclear model was a crucial step in understanding the structure of the atom

KEY THINKERS
Ernest Marsden (1889–1970)
Hans Geiger (1882–1945)
Ernest Rutherford (1871–1937)

WHAT COMES NEXT
Further understanding of the atom thanks to Chadwick's discovery of the neutron (see p.17)

SEE ALSO
Chadwick and the neutron p.17
The plum pudding model p.22
The Bohr model p.26

At the time of Rutherford's experiments, the prevailing theory regarding the structure of the **atom** was J J Thomson's **plum pudding model**. Rutherford's observation that some of the alpha particles passed straight through the foil was consistent with Thomson's theory, which suggested that the majority of the atom was nothing more than empty space. However, the fact that others were dramatically deflected did not match the plum pudding model. Rutherford was astonished and said, "It was quite the most incredible event that has ever happened to me in my life. It was almost as incredible as if you fired a 15-inch shell at a piece of tissue paper and it came back and hit you."

In the 1911 paper that summarized his findings, Rutherford explained the observations by proposing the **nuclear model**, in which a dense mass was localized in the atom, and that when some of the positive alpha particles encountered it, they had been repelled and had "bounced back" towards their origin. He assumed that the **nucleus** was positively charged, but it wasn't until 1913, and after further experiments, that he could confirm that. In 1920 Rutherford christened the positive fundamental particle the **proton**.

In 100 words

Between 1908 and 1911, Ernest Marsden and Hans Geiger performed several scattering experiments in the laboratory of Ernest Rutherford at the University of Manchester in England. In the Gold Foil Experiment, alpha particles (positively charged tiny particles) were fired at a thin piece of gold foil. Most of the alpha particles passed straight through the foil, but others were deflected at varying angles, with some even rebounding directly back towards the source. Rutherford concluded that the **atom** had a dense centre of mass, which he called the ***nucleus***, and with it, the **nuclear model** of the atom was conceived.

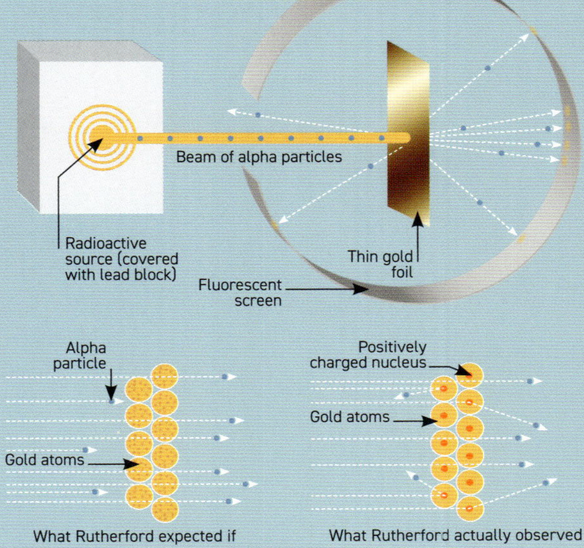

THE GOLD FOIL EXPERIMENT
Rutherford fired alpha particles at gold foil, and the angles that the particles deflected suggested the presence of the nucleus.

13

The Bohr model

WHY IT MATTERS
Bohr's model provided the bridge between a simple nuclear model and the beginning of incorporating quantum mechanics into atomic theory

KEY THINKERS
Niels Bohr (1885–1962)
Ernest Rutherford (1871–1937)

WHAT COMES NEXT
Quantum mechanics taking us to the modern understanding of the atom and its structure

SEE ALSO
The nuclear model p.24
The quantum mechanical model p.28

Bohr's **model** was relatively quickly made obsolete by the advent of true quantum mechanics, which accounted for the behaviour of more complicated **atoms**. However, the simplicity of the model and its fundamentally easy-to-understand ideas mean that it is still used to this day as a common tool when teaching atomic structure.

"Everything we call real is made of things that cannot be regarded as real. If quantum mechanics hasn't profoundly shocked you, you haven't understood it yet"

Niels Bohr, physicist (1958)

In 100 words

Following Rutherford's work in establishing the **nuclear model**, the Danish physicist Niels Bohr pushed atomic theory forward with his own model in 1913. Like Rutherford, Bohr also envisaged a small, dense **nucleus** but one surrounded by **electrons** orbiting in fixed, circular paths at specific distances from the nucleus. This model was groundbreaking, as it was first to incorporate quantum theory, explaining the discrete emission spectrum of hydrogen. The model worked well with a simple **atom** with a single electron, such as hydrogen, but was limited with more complex atoms. However, it paved the way for more sophisticated quantum mechanical models.

THE BOHR MODEL
Bohr's model of the atom suggested a small, dense nucleus surrounded by orbiting electrons.

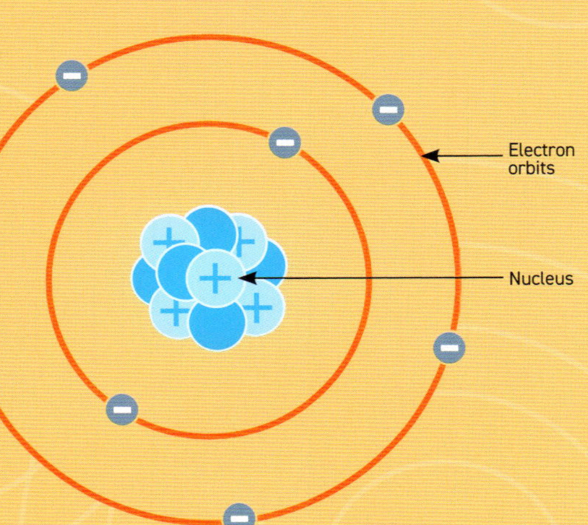

14

The quantum mechanical model

WHY IT MATTERS
The paradigm change, from electrons as particles to their duality as waves, formed the basis for all modern atomic theory

KEY THINKERS
Louis de Broglie (1892–1987)
Erwin Schrödinger (1887–1961)
Werner Heisenberg (1901–1976)
Max Born (1882–1970)

WHAT COMES NEXT
Further development of the modern understanding of the atom via leptons, quarks and bosons (see p.168)

SEE ALSO
The Bohr model p.26

Bohr's idea of quantized levels for **electrons**, together with de Broglie's wave-particle duality, represented a complete paradigm change in thinking regarding the electron, and in turn the ideas surrounding atomic theory. The wave-like nature of electrons opened a whole new move away from classical, Newtonian physics. Schrödinger calculated the quantum numbers that describe the whereabouts of an electron. Heisenberg's uncertainty principle accounted for some limitations, by suggesting that knowing both the momentum of an electron and its position have limits of accuracy, and Max Born took Schrödinger's wave-functions and interpreted them as three-dimensional probability "maps" of where electrons would most likely be found. Together, these scientists and others created our modern understanding of atomic theory.

In 100 words

In 1913, the Bohr **model** was a huge step forward in understanding the **atom**, but it failed to explain atoms that were more complicated than hydrogen, and it also had other shortcomings. In 1924, Louis de Broglie suggested that particles could exhibit wave-like behaviour. The wave-particle duality idea was then used by Erwin Schrödinger to make the crucial breakthrough in 1926. His famous equation treated **electrons** as waves rather than particles, and in the process introduced the idea of atomic orbitals. Further work by Werner Heisenberg and Max Born completed what we now know as the quantum mechanical model of the atom.

The Elements

Every literal "thing" that you see in the macroscopic, observable world is a combination of the atoms of the currently known 118 elements. The discovery and categorization of the elements is a story of both human endeavour and scientific discovery. It's also laden with controversy, skullduggery and confusion – just like real life!

The diversity of the chemical elements is reflected in their properties and uses. From the benign to the deadly, from the supremely useful to the currently useless, from the colourless, odourless gases to the bright, shiny, colourful solids, and from the ubiquitous to the ultra-rare, all are represented with the promise of even more to come. Organized and catalogued on the most iconic of all chemical images – the periodic table – the elements truly are "chemistry".

15

In 100 words

An **element** is a substance that cannot be broken down by a chemical reaction to produce anything simpler. A chemical reaction of an element can only produce something that is more complex than the element. For example, carbon can combine with oxygen to become carbon monoxide or carbon dioxide. Each element is characterized by the type of **atom** that it is made from. All neutral atoms of the element carbon contain a unique, fixed number of **protons**, along with the same number of **electrons**, plus varying numbers of **neutrons**. It is the unique number of protons that define each element.

The elements

Early Greek philosophers proposed that matter consisted of four elements: air, fire, water and earth, and that the amount of each that was present in a substance determined its properties. Democritus first suggested the idea of atomization – that matter consisted of tiny, indivisible particles. In the 18th century, chemistry was coming into its own as a science based upon quantifiable experimentation. Much work, by many people, notably Priestley and Lavoisier, was built upon by English chemist Dalton. In the early 19th century, Dalton published his classic paper *A New System of Chemical Philosophy*, which proposed the unique nature of each element's atoms and forms the basis of our modern understanding.

WHY IT MATTERS
The fundamental building blocks, the 118 elements combine to make up all other matter

KEY THINKERS
Empedocles (c.490–430 BCE)
Democritus (460–370 BCE)
Antoine Lavoisier (1743–1794)
John Dalton (1766–1844)

WHAT COMES NEXT
The periodic table's seventh period was completed by the official naming of element number 118 in 2016. The question has become: How many more elements can be synthesized?

SEE ALSO
The conservation of mass p.16
John Dalton's atomic theory p.18
The nuclear model p.24

16

The naming of the elements

WHY IT MATTERS
Once established, an element's name and symbol can be never changed. Having an element named after a place or person is considered to be one of chemistry's greatest honours

KEY THINKERS
IUPAC – The International Union of Pure and Applied Chemistry (founded in 1919)
Glenn Seaborg (1912–1999)

WHAT COMES NEXT
With the establishment of strict guidelines, some of the romance may have been removed from the nomenclature of the elements

SEE ALSO
The discovery of the elements p.34

The 1947 directive gave naming rights to the scientists who demonstrated the existence of a new **element**, subject to approval by an IUPAC committee. However, bickering between US and Soviet scientists during the Cold War about who had actually discovered elements 104 to 106, and therefore who had the right to propose names, led to a period of time where multiple names were floating around for the same elements. In 1994, attempting to resolve the problem, IUPAC proposed a solution that only served to create even more arguments. It wasn't until 1997 that things were finally resolved. The most recent additions to the **periodic table**, confirmed in 2016, followed the newest IUPAC guidelines for naming synthetic elements. The process now involves the confirmation of a discovery, the provisional assignment of a Greek name and a position on the periodic table and an invitation to the discoverers to submit a name and symbol for the new element, using either a mythological concept, a mineral, a country or place, a property or a scientist. This is then subject to a public review of the name and symbol before any approval is given. One example of a scientist being honoured in this way is Glenn Seaborg's 1997 recognition of element 106 (seaborgium).

In 100 words

The ancient **elements** took their names from various cultures, often drawing inspiration from astronomy, especially the Sun (gold) and the planets. There was no established way of naming a new element and just about anything went until 1947, when IUPAC intervened with a set of guidelines. Mythology, geography, the properties of the element itself or the name of the discoverer could be used. Between the 1960s and the 1990s, disputes, called the Transfermium Wars, arose over the discovery and naming rights for the elements 104 to 106, and IUPAC had to step in once more, this time with a heavier hand.

ELEMENT NAMES AND SYMBOLS
Periodic tables typically represent elements with their atomic number, atomic weight, their symbol and name.

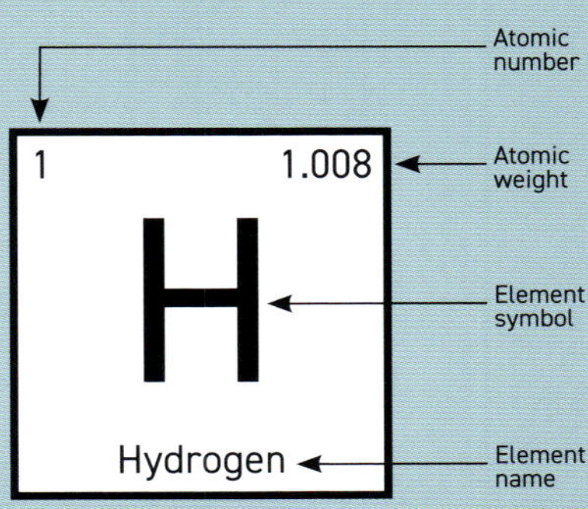

17

The discovery of the elements

WHY IT MATTERS
The fundamental building blocks of all chemical compounds, the elements make up all matter

KEY THINKER
Hennig Brand (c. 1630–c. 1692 or c. 1710)

WHAT COMES NEXT
Continued attempts to synthesize new elements to extend the periodic table beyond the current 118 known elements (see p.46)

SEE ALSO
The naming of the elements p.32

The reasons for such ambiguities are various, not least of all because many **elements** were discovered long before the formal recording of history and before a connected world existed. The prehistoric elements of antiquity include copper, gold, silver, lead, tin and mercury. Equally importantly, many elements were in use long before the definition was established for what an element actually is. Even after that, from the 17th to the 20th centuries, chemists across the world were not necessarily communicating with one another to any degree, nor was the dissemination of information as instantaneous as it is now. This led to many disputes, such as the one between Priestley and Scheele over the discovery of oxygen. It's generally accepted that Scheele probably did discover oxygen first, but his work wasn't published until after Priestley's and so the latter is given credit.

Even in the modern world, there have been many disputes, especially over the discovery of newer, synthesized superheavy elements (see p.32). In the 1950s, driven in part by the Cold War, there was an enormous argument between Soviet and American scientists (and for a short while, Swedish scientists) about the discovery of element 102 (nobelium). There are several examples of such wrangling and even the complete fabrication of data, such as when Bulgarian chemist Victor Ninov was found guilty of falsifying data about discovering elements 116 and 118.

In 100 words

The first **element** to be discovered by and clearly attributed to a single person was phosphorus in 1669. German alchemist Hennig Brand stumbled across it while boiling urine in his laboratory in his search for the philosopher's stone (see p.10). The attribution of discoveries are difficult in some cases, impossible in others and full of disputes in many! Some might argue that arsenic, not phosphorus, is the first modern element that can be traced to a single person, citing German philosopher Albertus Magnus's isolation of it in 1250, but to add to the confusion, isolation is not necessarily discovery.

18

Allotropes

WHY IT MATTERS
The properties of different forms of exactly the same element can vary widely

KEY THINKER
Jöns Jacob Berzelius (1779–1848)

WHAT COMES NEXT
The discovery of fullerenes in the 1980s was a surprise. More allotropes may be waiting to be discovered

SEE ALSO
The elements p.31
Fullerenes p.38

The carbon **atoms** in diamond are arranged in tetrahedrons, in which each atom is covalently bonded (see below right) to four others. The pattern is repeated trillions and trillions of times within a tiny piece of diamond. The **covalent bonds** are all extremely strong, making diamond an incredibly hard substance with a high melting point. In graphite, the carbon atoms are bonded to one another differently. Here, each one is bonded to only three others to give a flat, honeycomb-like pattern arranged in layers. These layers are attracted to one another by only weak dispersion forces (see p.76), allowing them to slide over one another easily. This gives graphite a slippery feel, allows it to act as a lubricant, and causes it to be much softer than diamond. How the carbon atoms are bonded makes all the difference.

Elements besides carbon can also be found as **allotropes**. For example, phosphorus can exist in different forms that are distinguished by their colours – white, red, violet and black. Arsenic is another element with multiple allotropes, also categorized by colours, this time yellow, grey and black. Swedish super-chemist Jöns Jacob Berzelius was the first to use the term *allotrope* in 1841.

ALLOTROPES OF CARBON
Graphite and diamond are allotropes of carbon. They are both made of carbon atoms, but the atoms are arranged in different ways.

In 100 words

Allotropes are different forms of the same **element**. An example of allotropy can be seen in two very familiar substances: diamond and graphite. These are both composed of carbon **atoms**, so how can it be that one of the sparkliest, hardest and most valued substances on Earth can be chemically the same as a soft, black, almost worthless substance used as pencil "lead"? The answer is found within the substances' structures. How the atoms are arranged and bonded together in any given allotrope makes a huge difference to its properties.

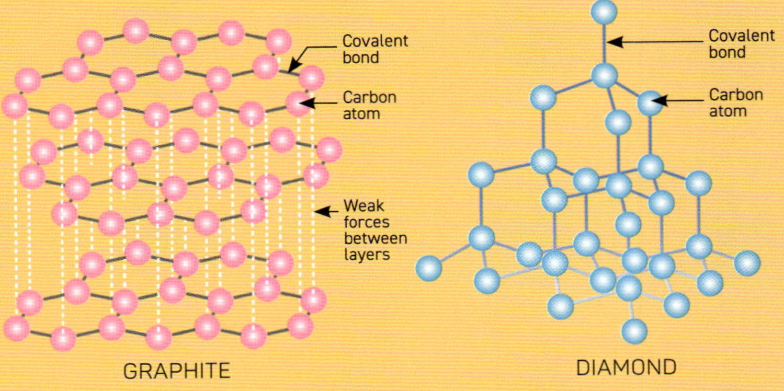

19

Fullerenes

WHY IT MATTERS
The unique structures of fullerenes make them potentially useful in areas such as medicine and electronics

KEY THINKERS
Richard Smalley (1943–2005)
Robert Floyd Curl Jr (1933–2022)
Harry Kroto (1939–2016)

WHAT COMES NEXT
Carbon nanotubes are fullerenes that represent an exciting look into the future of nanotechnology

SEE ALSO
Allotropes p.36
Nanotechnology p.170

Discovered in the 1980s, **fullerenes** are **allotropes** of carbon bound together by **covalent bonds**. The most famous of these is buckminsterfullerene, which has a cage-like structure resembling a football that is made up of pentagons and hexagons of carbon **atoms**.

As a group of allotropes, fullerenes contain not only the classic "soccer ball" C_{60} but also others with 20, 70 and 540 carbon atoms joined together in various 3D shapes. A lot of research has gone into using fullerenes as hydrogen storage vehicles. The idea is that the hollow, cage-like structures can be used as carriers for hydrogen **molecules**, so the hydrogen can be stored and transported for use as an alternative energy source.

"Scientific discoveries matter much more when they're communicated simply and well – if you can't explain your work to the man in the pub, what's the point?"

Harry Kroto, chemist (1998)

In 100 words

Fullerenes are in part interesting because they were discovered so relatively recently. It wasn't until the 1980s, when Harry Kroto, Richard Smalley and Robert Curl Jr started using a laser to investigate carbon, that they discovered that **atoms** were clumping together into single entities as they cooled. The structure they found most often was a collection of 60 carbon atoms, bound together in rings of five and six carbons, creating a familiar, football-shaped 3D structure. These new structures were named buckminsterfullerene after the geodiscs that had been popularized by the American architect Richard Buckminster Fuller.

BUCKMINSTER-FULLERENE
A fullerene with the formula C_{60} has a fused, cage structure resembling a football.

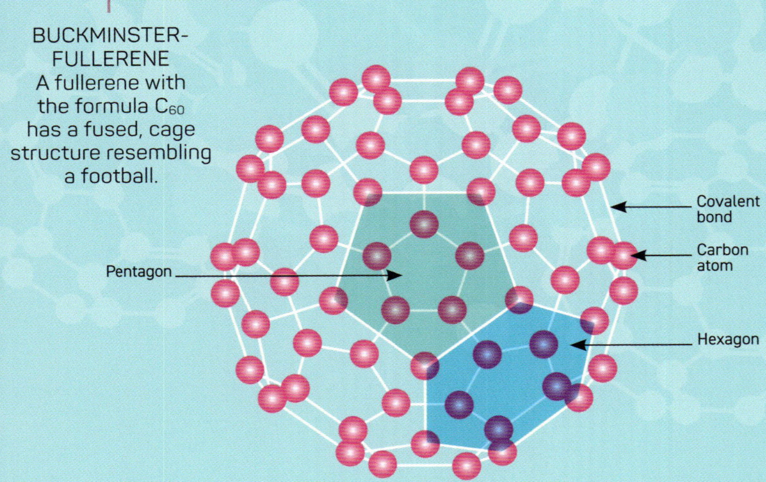

BUCKMINSTERFULLERENE

Periodic pioneers

WHY IT MATTERS
Long before Mendeleev's pivotal work in 1869 (see p.42), several other chemists attempted to bring the elements to order

KEY THINKERS
Johann Wolfgang Döbereiner (1780–1849)
Alexandre-Émile Béguyer de Chancourtois (1820–1886)
John Newlands (1837–1898)
Julius Lothar Meyer (1830–1895)
Henry Moseley (1887–1915)

WHAT COMES NEXT
Most of the work discussed here predates Mendeleev's 1869 work, which became the seminal work for modern periodicity

SEE ALSO
Mendeleev p.42
The modern periodic table p.46

Mendeleev's original organization of the **elements** was based upon **atomic weights**. In that regard, his work would never have been possible without the contribution of Lavoisier's law of conservation of mass and subsequent work on mass by John Dalton and others. Our modern table is organized by **atomic numbers**, largely thanks to the work of Henry Moseley. In 1913, working with X-rays and the elements known at the time, Moseley's work gave credence to the idea of an ascending atomic number that matched the known patterns of chemical properties of the day. This removed the need for the occasional fudging of order that was sometimes required when listing according to atomic weight.

Before 1869, other scientists had contributed greatly to the organization of the elements. Döbereiner's triads were groups of three elements, organized by atomic weight, for which he demonstrated that the middle element had a weight of approximately the average of the other two, and that its properties were also intermediate. Béguyer de Chancourtois's telluric screw was an elaborate, three-dimensional cylinder that allowed for the vertical alignment of similar elements by rotating the cylinder. In 1865, Newland's law of octaves noted that periodic patterns were observable every seven elements. In 1868, just a few months before Mendeleev's presentation to the Russian Chemical Society, Meyer had proposed a remarkably similar organization, but it wasn't published until 1870. With Meyer working in Germany, and Mendeleev in Russia, it was entirely possible for the two men to reach almost identical conclusions simultaneously and entirely independently.

In 100 words

It's hard not to singularly reference Mendeleev when it comes to the origins of the **periodic table**, but the contributions of many others deserve credit too. From Döbereiner's triads in 1829 and Béguyer de Chancourtois's telluric screw in 1862, to Newland's law of octaves in 1865 and Meyer's classification of **elements** between 1868 and 1870, the Russian father of the periodic table was far from the only thinker on the organization of the elements. After Mendeleev's death, Henry Moseley's work on **atomic numbers** was also instrumental in forming the table as we know it today.

DÖBEREINER'S TRIADS
These are groups of three elements, organized by atomic weight, where the middle element's weight is approximately the average of the other two.

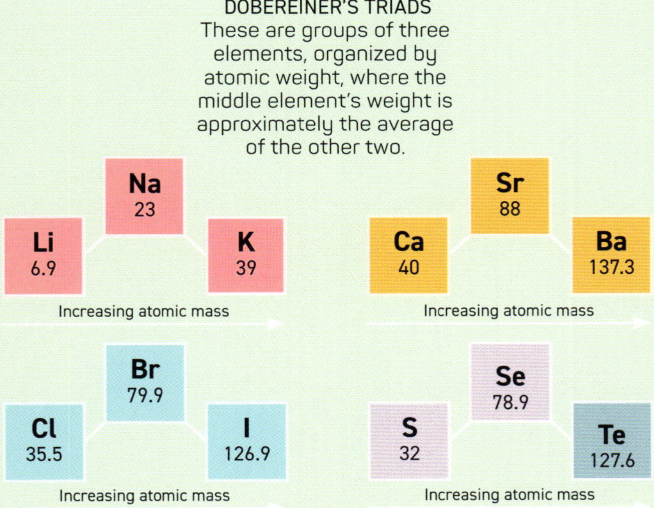

21

Mendeleev

WHY IT MATTERS
The father of the modern periodic table

KEY THINKER
Dmitri Mendeleev (1834–1907)

WHAT COMES NEXT
The possibility of more elements beyond those in the current periodic table

SEE ALSO
Periodic pioneers p.40
The modern periodic table p.46

Dmitri Mendeleev was born in Siberia in 1834. His early life was hard but also typical by mid-19th century Siberian standards. He was one of at least 14 children and the family – including himself – suffered poor health, and his parents died relatively early in his life. However, before her death, his mother had recognized Dmitri's academic potential, and she encouraged his studies. Mendeleev worked briefly as a teacher in Crimea, in 1855, and after the outbreak of the Crimean War he travelled to Odessa to work. After stints as a professor in St Petersburg and time spent in Paris and Heidelberg, Mendeleev returned to St Petersburg, ultimately becoming the chair of inorganic chemistry at the university there. It was while he was in that position that he wrote a number of textbooks, including *The Principles of Chemistry*. He identified a need for more up-to-date references. His work in this area was important for the advancement of chemistry in Russia and was central to the development of his periodic law.

In terms of the **periodic table**, Mendeleev's genius was twofold. Firstly, he was flexible enough to realize that when the **atomic weights** of known **elements** didn't quite fit the expected order, he could move them around to accommodate them. For example, the positions of iodine and tellurium, in terms of atomic weights, are reversed on the table. Secondly, he left spaces for as yet undiscovered elements. His predictions regarding the properties of those missing elements were astonishingly accurate.

In 100 words

When, in early 1869, Mendeleev presented his thoughts about the relationship between the properties and **atomic weights** of the **elements** to the Russian Chemical Society, he could not have known the mark that he was about to leave on both chemistry and the larger culture as a whole. As established earlier (see p.40), many ideas regarding the classification, organization and grouping of elements preceded his work, but Mendeleev's creation became the salient one, from which all modern **chemical periodicity** grew, thanks to its comprehensive and enduring nature.

22

Element symbols

WHY IT MATTERS

In order to clearly and consistently communicate chemical ideas, establishing a standardized system of nomenclature was essential

KEY THINKERS

John Dalton (1766–1844)
Jöns Jacob Berzelius (1779–1848)

WHAT COMES NEXT

With the establishment of formulae and the conservation of mass, full chemical equations follow naturally

SEE ALSO

The elements p.31
The naming of the elements p.32
The discovery of the elements p.34
Chemical equations p.75

The earliest practitioners in fields that would ultimately become modern chemistry were the Greeks and the alchemists. They used symbolic representations to depict their ideas, mostly in the form of shapes such as triangles of various orientations. The alchemists developed the symbol idea with more complex shapes, often based upon planetary and astrological symbols. In the early 19th century, John Dalton continued the practice by publishing a variety of circular symbols that used a combination of shapes and letters to represent the known **elements** of the time. The modern system, where the letter symbols are drawn from the element names, was proposed by Swedish chemist Jöns Jacob Berzelius in 1814.

In 100 words

The **element** symbols are composed of either a single or a two-letter depiction. Single letters are capitalized, and two-letter symbols have capitalized first letters followed by a lower-case second. The element symbols are combined, along with subscript integers, to create the chemical formulae that show the make-up of **compounds**. For example, the element symbols for sulphur, oxygen and chlorine are S, O and Cl respectively, and one of the compounds that they combine to make is thionyl chloride, $SOCl_2$. Each formula unit of thionyl chloride contains one sulphur **atom**, one oxygen atom and two chlorine atoms.

23

The modern periodic table

WHY IT MATTERS
Iconic and of practical importance, the periodic table is a predictive tool that is fundamental to all chemistry

KEY THINKERS
Dmitri Mendeleev (1834–1907)
Glenn Seaborg (1912–1999)

WHAT COMES NEXT
How many new elements can be synthesized? Does the periodic table have a natural "end"? Will it need to be rearranged in the future?

SEE ALSO
The elements p.31
Periodic pioneers p.40
Mendeleev p.42
Chemical periodicity p.48

An iconic symbol of 20th- and 21st-century life, the reach of the modern **periodic table** goes far beyond chemistry and science, outside of the classroom and into the wider domain of the culture as a whole. Its potential for bringing organization and uniformity to otherwise apparently disparate objects or ideas has been exploited by countless people. Think of something that you'd like organized and classified into *The Periodic Table of* [...] [insert your chosen "thing"], and it's probably already been done; The Periodic Table of Elephants, anyone? Or perhaps The Periodic Table of Sandwiches? Yep, both have already been concocted.

Within chemistry itself, the role of the table in allowing scientists to make predictions about the behaviour of **elements** is crucial. Its development over time is a fascinating one of history, science, experimentation, skulduggery and subterfuge, and it remains a supreme symbol of human endeavour.

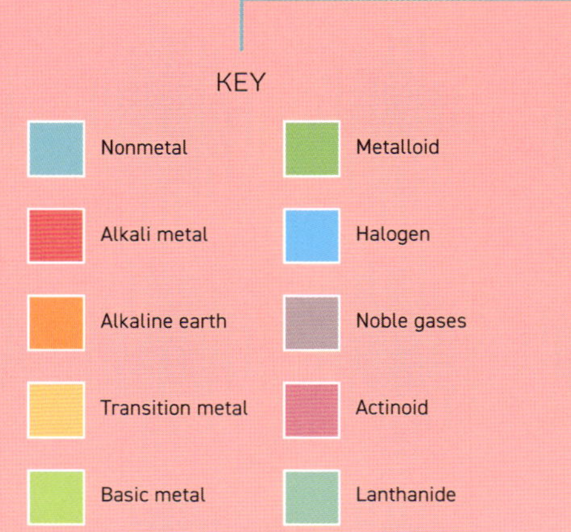

In 100 words

The **periodic table** collects the 118 currently known chemical **elements**, organizes them, and in the process creates one of the most important tools in chemistry. Arranging the elements by increasing **atomic number** (the number of **protons** in a single **atom** of each unique element), it places chemically similar elements in vertical columns, known as groups, and elements with gradually changing characteristics in horizontal rows, called periods. The relative position of each element allows predictions to be made about its behaviour and properties. Metals are found predominantly on the left and centre of the table, with nonmetals in the top right.

MODERN PERIODIC TABLE OF ELEMENTS

1 IA	2 IIA	3 IIIB	4 IVB	5 VB	6 VIB	7 VIIB	8 VIIIB	9 VIIIB	10 VIIIB	11 IB	12 IIB	13 IIIA	14 IVA	15 VA	16 VIA	17 VIIA	18 VIIIA
H																	He
Li	Be											B	C	N	O	F	Ne
Na	Mg											Al	Si	P	S	Cl	Ar
K	Ca	Sc	Ti	V	Cr	Mn	Fe	Co	Ni	Cu	Zn	Ga	Ge	As	Se	Br	Kr
Rb	Sr	Y	Zr	Nb	Mo	Tc	Ru	Rh	Pd	Ag	Cd	In	Sn	Sb	Te	I	Xe
Cs	Ba	57-71 Lanthanides	Hf	Ta	W	Re	Os	Ir	Pt	Au	Hg	Tl	Pb	Bi	Po	At	Rn
Fr	Ra	89-103 Actinides	Rf	Db	Sg	Bh	Hs	Mt	Ds	Rg	Cn	Nh	Fl	Mc	Lv	Ts	Og

La	Ce	Pr	Nd	Pm	Sm	Eu	Gd	Tb	Dy	Ho	Er	Tm	Yb	Lu
Ac	Th	Pa	U	Np	Pu	Am	Cm	Bk	Cf	Es	Fm	Md	No	Lr

24

Chemical periodicity

WHY IT MATTERS
The ability to predict the properties of the elements and their compounds, based only upon patterns, is a vital tool in the chemist's arsenal

KEY THINKER
John Newlands (1837–1898)

WHAT COMES NEXT
As new elements are created, chemists can use periodicity to reasonably predict properties that might otherwise be impossible to observe

SEE ALSO
Periodic pioneers p.40
Mendeleev p.42
The modern periodic table p.46

If you understand musical notes, then you can understand **periodicity.** The musical notes A to G can be thought of as a left-to-right period on the **periodic table.** Moving down an octave can be thought of moving down to the next period (row). The A to G notes in this new period are different to the ones in the first, but each A is related to the previous A, each B is related to other Bs, etc., and there is a predictable pattern from A to G across the all periods. **Elements** work in exactly the same way. Because we know something about the elements in the second period (Li, Be, B, C, N, O, F and Ne), we can make very accurate predictions about the properties of elements in the third period (Na, Mg, Al, Si, P, S, Cl and Ar). There is a repeatable pattern that we call *The Periodic Law.* The first scientist to champion the idea of the elements having an analogous relationship with the periodic patterns of musical notes was John Newlands. His paper, *The Law of Octaves, and the Causes of Numerical Relations among the Atomic Weights*, was published in 1865.

Important periodic properties include atomic radius and ionic radius (the size of the **atoms** and **ions**), first ionization energy (the energy required to remove an **electron** from an atom) and **electronegativity** (see p.66), which change very predictably across periods and down groups. The ideas behind periodicity allowed Mendeleev to predict the future discovery of some elements, and they allow modern chemists to do the same regarding new elements.

PERIODICITY
Many chemical and physical properties of the elements vary somewhat uniformly and consistently with their position in the periodic table.

In
100
words

Periodicity is the observation that the properties of the **elements** and their **compounds** can often be accurately predicted by the relative positions of the elements in the **periodic table**. The periodic table is arranged in rows (periods) and columns (groups), and periodicity relies upon the repeating patterns that are found as the table is traversed across periods and when moving up and down groups. Elements in the same group tend to exhibit similar chemical properties due to their outer (valence) electronic configurations being similar. Traversing a period results in gradual changes in properties, whose characteristics are repeated in subsequent rows.

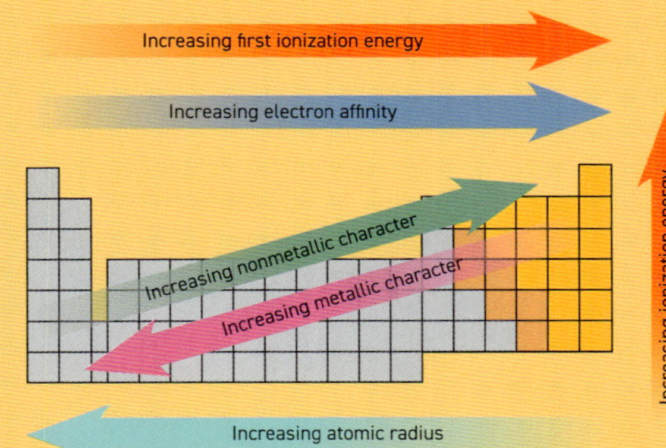

25

The alkali metals

WHY IT MATTERS
Group 1 elements play important roles, including sodium and potassium in the body's nervous system and lithium in modern battery technology

KEY THINKER
Humphry Davy (1778–1829)

WHAT COMES NEXT
An increasing demand for lithium because of electric vehicles increases pressure on the element's supply

SEE ALSO
The modern periodic table p.46
The alkaline earth metals p.52
Electrolytic cells p.128

In some ways, the alkali metals are not typical metals. For example, in addition to their reactivity, they have relatively low melting points – not a property typically associated with metals. Because of their reactivity, they are only found in nature in **compounds** where they have already combined with other **elements**. The family name of "alkali" for group 1 came about because their oxides (compounds with oxygen), will dissolve in water to produce a basic (or "alkaline") **solution**. On some **periodic tables** you may see hydrogen included at the top of group 1, as either an integral or an associated member of the group, but the only reason for it being included is that it has an outer (valence) electronic configuration (s1), which is also the case for the other members of the group. That is where the similarities begin and (mostly) end.

English chemistry pioneer Humphry Davy is closely associated with the alkali metals. He used electrolysis to decompose compounds into their elements, and in 1807, he isolated potassium and sodium from molten potash and soda respectively. He also succeeded in isolating magnesium, calcium, strontium, barium and boron.

In 100 words

The group 1 **elements** are found on the far left of the **periodic table**, in the very first column. They include lithium (Li), sodium (Na), potassium (K), rubidium (Rb), cesium (Cs) and francium (Fr). These elements are all reactive metals; in most cases they react violently with water to produce hydrogen gas, and they tarnish quickly with the oxygen in the air to create a dull, oxidized surface on the metal. Other vigorous reactions with the halogens can also be observed that lead to more stable **compounds**, such as sodium chloride. The metals themselves are soft, and they can be easily cut with a knife.

GROUP 1 ELEMENTS
Group 1 elements, found on the far left of the periodic table, are all reactive metals.

The alkaline earth metals

Group 2 **elements** are more typically metallic than group 1 elements, being harder and having higher melting points than their counterparts to the left. They are linked by their outer electronic configurations all ending in s2, meaning that they exhibit similar properties to one another, one of which is their tendency to form 2+ **ions** in ionic **compounds** (see p.68).

In addition to his work with group 1 elements, Davy was also the first to isolate magnesium, calcium, strontium and barium. His work in electrolysis, via the discovery of several group 1 and group 2 elements, was truly pioneering.

WHY IT MATTERS
Calcium and magnesium are important in biological functions, including bone growth

KEY THINKER
Humphry Davy (1778–1829)

WHAT COMES NEXT
Magnesium is being considered as an alternative to lithium in battery technology

SEE ALSO
The alkali metals p.50

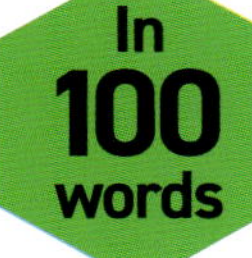

Six elements make up the alkaline earths: beryllium (Be), magnesium (Mg), calcium (Ca), strontium (Sr), barium (Ba) and radium (Ra). Like group 1, group 2 is a collection of relatively reactive metals. They are generally less "explosive" than their neighbours, but they still react quickly and tarnish easily by reacting with the oxygen in the air. That reaction is linked to their collective name. When originally discovered, many were found in nature, in the Earth, as their oxides. Even after it was found that these "earths" were not pure elements, the name still persisted for the elements themselves.

In 100 words

Most commonly known as "the boron group", group 13 consists of the **elements** boron (B), aluminium (Al), gallium (Ga), indium (In), thallium (Tl) and nihonium (Nh). Like several other p-block groups, this group contains elements with diverse properties, with the metal aluminium being the most familiar. Aluminium is the most abundant metal in Earth's crust, and its widespread use is centred around it being both very strong and very light. Its resistance to significant corrosion adds to its allure. The metals gallium and indium are less well-known but have some specialized uses, for example indium is ubiquitous in touchscreens.

The group 13 elements

Boron is a metalloid (see p.46). In school chemistry laboratories there is a lot of boron, although you might not realize it. Borosilicate glass is used to make lab glassware since it is heat-resistant and strong. Aluminium is often used in high-performance vehicles like aircraft and space vehicles because of its high strength-to-weight ratio. Aluminium was once considered a precious metal since it was hard to extract from its ores. This all changed when Hall and Héroult perfected the electrolysis process for obtaining the metal.

Gallium is found in semiconductors and indium in all touchscreens. Thallium is a highly toxic element, which has poisoned several people both accidentally and by design! Nihonium is the newest group 13 element, named only in 2016 after the Japanese name for Japan (Nihon), where it was first synthesized by a team led by Kōsuke Morita. Very little is known about this **radioactive** element.

27

WHY IT MATTERS
Group 13 contains a wide variety of elements with an equally large range of real life applications

KEY THINKERS
Kōsuke Morita (1957–)
Charles Martin Hall (1863–1914)
Paul Héroult (1863–1914)

WHAT COMES NEXT
Gallium is a famously low-melting point metal that can be easily moulded and shaped, meaning it has an exciting future in the electronics industry

SEE ALSO
Electrolytic cells p.128

28

The group 14 elements

WHY IT MATTERS
Some of the most important elements known congregate in group 14

KEY THINKER
Georgy Flerov (1913–1990)

WHAT COMES NEXT
The use of carbon nanotubes and graphene represent exciting future uses of carbon in the realm of nanotechnology

SEE ALSO
The group 13 elements p.53
The pnictogens p.55
Organic chemistry p.136

Elements in the same group often have similar properties, but group 14 is an exception. In the classification of the elements we see a rare diversity. At the top is nonmetal carbon, followed by the silicon and germanium metalloids, then two classic metals in tin and lead, and finally a newly synthesized element that we know virtually nothing about, flerovium. Flerovium was named after Georgy Flerov, as was the Laboratory of Nuclear Reactions in Dubna, Russia, where the element was first discovered.

Carbon, tin and lead are all elements of antiquity. Tin is part of the alloy bronze, which established the era known as the Bronze Age. Carbon's use in nanotubes and other modern **allotropes** (see p.36), along with silicon and germanium semiconductors, contribute to modern chemistry.

In 100 words

Found in the p-block on the **periodic table**, group 14 contains a diverse collection of elements. Known simply as "the carbon group", it has also been called the tetrels, the crystallogens and the adamantogens. The group contains carbon (C), silicon (Si), germanium (Ge), tin (Sn), lead (Pb) and flerovium (Fl), which range from nonmetals to metals, with a metalloid in the mix too. The group certainly has a claim to being the most important group of the periodic table since it contains some of the most vital elements, with carbon at the centre of all lifeforms, and silicon at the centre of computing.

In 100 words

Six **elements** are found in group 15 of the **periodic table**, and they are collectively known as the pnictogens. The weird name for the group comes from the Greek *pnikta* meaning "to choke" or "to suffocate", and it refers to nitrogen's inability to support combustion and respiration (breathing). Beyond nitrogen (N), we find phosphorus (P), arsenic (As), antimony (Sb), bismuth (Bi) and moscovium (Mc). Metals, nonmetals and metalloids are all represented, with phosphorus and arsenic existing as a number of **allotropes** (see p.36). With the exception of unreactive nitrogen gas – which makes up approximately 78 per cent of the atmosphere around us – all of the other elements are solids.

29

The pnictogens

The pnictogens have many uses. Nitrogen is a crucial nutrient for plants, and as such, it is a component of many fertilizers (see p.112). Phosphorus is usually attributed as the first element to be discovered by an identifiable person – the alchemist Hennig Brand in 1669 (see p.35). It is a fearsome element that can inflict awful burns and has been used in chemical warfare, but when controlled, it can be used in match heads. Phosphorus also has a vital role to play in human biology, in bone and teeth enamel and in RNA and **DNA** (see p.164). Arsenic has been used as a medicine in the past, but it is also known to be a potent poison. Bismuth treats stomach upsets. **Compounds** containing antimony have been used as black eye liners for thousands of years, often known as *kohl*. The group is rounded out by one of the newest elements, moscovium. It was named after Moscow for the contribution of Russian scientists, led by Yuri Oganessian, who also had element 118 named after him.

WHY IT MATTERS
Group 15 contains a wide variety of important elements that have a rich history – both good and bad!

KEY THINKERS
Hennig Brand (c. 1630–c. 1692 or c. 1710)
Fritz Haber (1868–1934)
Yuri Oganessian (1933–)

WHAT COMES NEXT
Antimony is a key component of many liquid metal batteries (LMBs) that may have a significant role to play in future alternate energy production

SEE ALSO
Allotropes p.36
The group 14 elements p.54
The chalcogens p.56
The Haber-Bosch process p.112

The chalcogens

In one way, we are happy that oxygen makes up so much of our atmosphere, since we have to breathe, but at the same time, its presence can be a serious problem. Over time, oxygen gas will react with anything that it comes into contact with. The most common problem is when it reacts with metals. It can discolour silver cutlery to give a dull, blackish coating, but it is even more problematic when the metal is iron. The **oxidation** of iron by oxygen in the presence of water is better known as rusting, and it is a real problem in the world everywhere. Oxygen is the most abundant **element** on Earth and the third most in the universe.

There are many interpretations of the origins and meaning of the term *chalcogen*; the most likely appears to be "ore former". Its legitimacy is bolstered by the fact that virtually all-important metal ores are either oxides or sulphides.

WHY IT MATTERS
Oxygen, the gas essential to respiration and life, heads group 16

KEY THINKERS
Carl Wilhelm Scheele (1742–1786)
Joseph Priestley (1733–1804)

WHAT COMES NEXT
Tellurium's use in cadmium telluride thin-film solar panels is important to future energy production

SEE ALSO
The pnictogens p.55
The halogens p.57

In 100 words

As with many groups in the p-block of the **periodic table**, we find a real mixed bag in group 16. Leading off is perhaps the most familiar element of all, oxygen (O). Making up approximately 21 per cent of the air around us, it is of course essential for life. Next, we find sulphur (S), a nonmetal like oxygen, and selenium (Se) and tellurium (Te), two metalloids (which have properties of both metals and nonmetals). Finally, two **radioactive** metals round out the chalcogens: polonium (Po) and one of the newest elements, livermorium (Lv), about which almost nothing is known.

In 100 words

Group 17 of the **periodic table** is a collection of reactive and diverse **elements**. Among fluorine (F), chlorine (Cl), bromine (Br), iodine (I), astatine (At) and tennessine (Ts), you'll find elements that are gases, a liquid and a solid at room temperature. Two of them are extraordinarily rare, **radioactive** elements. The word halogen means "salt forming", which means that the group 17 elements will react with metals to form **compounds** that we call salts. Salts are compounds with positive **ions** (cations) and negative ions (anions) combined together. In each salt, the halogens provide the negative halide like fluoride (F^-), chloride (Cl^-), bromide (Br^-) and iodide (I^-).

31

The halogens

Carrying on his expertise in all matters of chemical nomenclature and classification, Jöns Jacob Berzelius first proposed the term *halogen* in 1826. The halogens can be a seemingly contradictory bunch. Fluorine is a dangerous, poisonous and pale-yellow gas that was responsible for killing several of the early chemists. Henri Moissan is notable for taming the element and that feat was – in part – sufficient for him to be awarded the 1906 Nobel Prize in Chemistry. In the form of fluoride ions, it has been used in water supplies and in toothpaste to help prevent tooth decay. Chlorine exhibits a similar pattern: it is a toxic gas that was used in World War I as a chemical weapon, but it is also widely used to make our water safe by killing bacteria. Chlorine, bromine and iodine are all plentiful in nature, being found in the oceans and in seaweed, but astatine and tennessine are two of the rarest elements on the whole periodic table.

WHY IT MATTERS
Fluorine, chlorine, bromine and iodine have important industrial, medical and environmental roles

KEY THINKERS
Henri Moissan (1852–1907)
Jöns Jacob Berzelius (1779–1848)

WHAT COMES NEXT
The persistence of some halogenated compounds in the environment, notably CFCs, have put their use under increasing scrutiny

SEE ALSO
The chalcogens p.56
The Haber-Bosch process p.112

32

The transition elements

WHY IT MATTERS
Transition elements are hugely important elements in industry as catalysts and in specialized alloys

KEY THINKERS
Charles Rugeley Bury (1890–1968)
William Hyde Wollaston (1766–1828)
Smithson Tennant (1761–1815)

WHAT COMES NEXT
The applications of transition metals are constantly evolving, but their supply for the short- and mid-term future remains a concern

SEE ALSO
Catalysts p.104

The transition **elements** are metals whose **atoms** have incomplete d sub-shells, which means they can form many **ions**. Unfortunately, this excludes elements such as scandium, zinc and cadmium that would otherwise be included when considering alternative definitions. So, there is a lot of ambiguity over what actually counts as a transition metal.

Regardless of the definition, these elements find use in a myriad of places, from the macro to the nano, in construction and technology and in traditional and innovative cutting-edge applications. They encompass the familiar (iron and copper), the ultra-rare (technetium), the exotic and expensive (rhodium and iridium), the ancient (gold and silver) and the modern and synthetic (meitnerium and copernicium). An eclectic bunch, they are both ubiquitous and invisible in everyday life, often appearing combined together. This allows the useful properties of the individual elements to be integrated to create even more useful alloys that have the benefits of multiple elements.

As a result of their extensive use, the majority of them are considered to be endangered elements – that is, elements whose supply is either under serious threat or increasing pressure.

There has been much debate as to how to define the transition **elements** as a collective, but the easiest way is to say that they are the elements found in groups 3 to 12 and in periods 4 to 7 of the periodic table. They are sometimes called the d-block elements because in this part of the **periodic table**, **electrons** are placed into d orbitals. They tend to be malleable, lustrous metals with a high melting point. The elements exhibit multiple **oxidation** states and form complex **ions**. They have a colossal number of special applications, including as **catalysts**.

THE TRANSITION ELEMENTS
Groups 3 to 12 and periods 4 to 7 contain the transition elements.

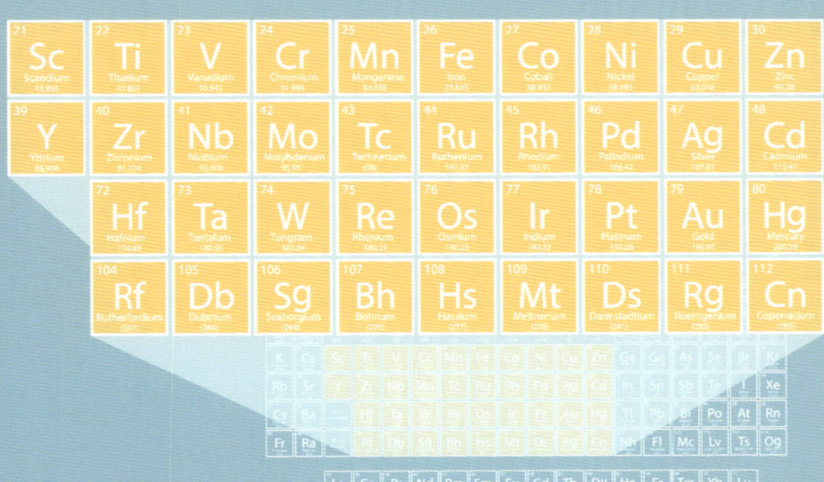

33

The noble gases

WHY IT MATTERS
Beyond their use in visible applications like neon signs, these elements are used in many other areas, such as superconductivity, quantum mechanics and general chemical reactivity

KEY THINKERS
Lord Rayleigh (1842–1919)
Sir William Ramsay (1852–1916)
Morris William Travers (1872–1961)

WHAT COMES NEXT
Originally thought to be almost completely inert, the group 18 elements have since been used to synthesize several compounds

SEE ALSO
The modern periodic table p.46

The **noble gases**, also known as the inert gases, are so called because for a long period of history they remained aloof. Their relatively unreactive nature made them difficult to detect, and their discovery eluded chemists until the latter part of the 19th century. It wasn't until argon was discovered by Lord Rayleigh and William Ramsay in 1894 that the ball got rolling. The discovery of helium soon followed in 1895. Ramsay isolated it from uranium minerals, after it had first been observed in a solar spectrum during an eclipse in 1868 by French astronomer Pierre Jules César Janssen. In 1898, Ramsay and Morris Travers also isolated krypton, neon and xenon from liquid air, and in 1900, the German Friedrich Ernst Dorn discovered radon. As such, Mendeleev's 1869 table couldn't have included any of these **elements** because none of them had been discovered at the time. When Mendeleev assembled his table, he left spaces for some elements, such as gallium, scandium and germanium. However, he didn't consider spaces for any of the elements found in group 18 since they didn't appear to be missing. The unique thing about these elements was that they formed a whole new group, and they were tacked on to the end of Mendeleev's table as a completely new addition.

In 100 words

Group 18 of the **periodic table** is the final column of the table, found on its far right-hand side. It contains seven **elements**: helium (He), neon (Ne), argon (Ar), krypton (Kr), xenon (Xe), radon (Rn) and oganesson (Og). The first six elements of the group are collectively known as the **noble gases**. These are naturally occurring, unreactive gases, but the final member of group 18 is not a gas at all. In fact, almost nothing is known of its chemistry since only a few, short-lived **atoms** have ever been synthesized. Current theory suggests that oganesson would most likely be a reactive solid.

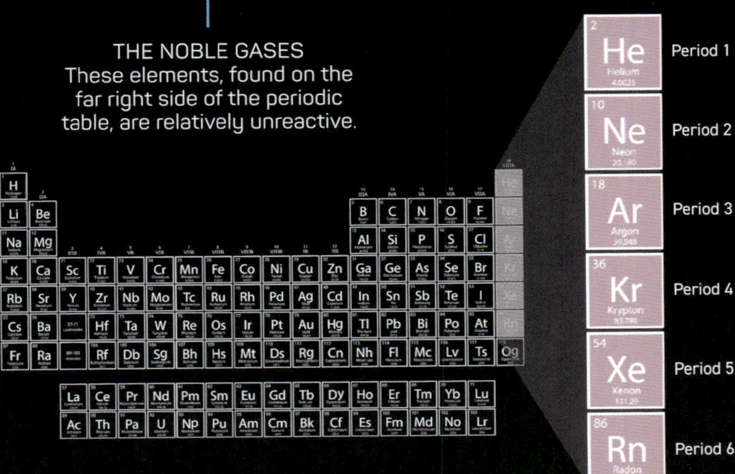

THE NOBLE GASES
These elements, found on the far right side of the periodic table, are relatively unreactive.

34

The actinides

WHY IT MATTERS
It's hard to overstate the significance of the nuclear age, and the profound impact of uranium and plutonium on history via nuclear weapons and power

KEY THINKERS
Enrico Fermi (1901–1954)
Glenn Seaborg (1912–1999)
Edwin Mattison McMillan (1907–1991)

WHAT COMES NEXT
New applications for the actinides in fields such as medicine and infinite industrial applications

SEE ALSO
The lanthanides p.64
Radioactivity p.130

Historically, the actinides fall largely into two distinct groups. Those discovered prior to 1940 with **atomic numbers** of 92 (uranium) and below, and those that have atomic numbers above 92 and were only produced from 1940 onwards. The second group is known as the transuranium **elements**. The split reflects the fact that, with only a few exceptions, the elements from hydrogen (atomic number 1) to uranium (atomic number 92) have stable **isotopes** (see p.20) that can be found in nature. Although two of the elements with atomic numbers greater than 92 (neptunium and plutonium) can be found in trace amounts on Earth, the others cannot, and as such they had to be born out of the nuclear age.

Because of their rarity and **radioactivity**, most actinides have very few uses outside some specialized applications, such as the use of americium in smoke detectors. Human-made, they are produced in only very small quantities and are used mostly for research. The notable exception is plutonium, which is used extensively in nuclear power plants, in nuclear weapons and in space exploration.

In 100 words

The **elements** with **atomic numbers** 89 (actinium) to 103 (lawrencium) are collectively known as the actinides. The name is derived from the first element in the series. Found at the very bottom of the **periodic table** in the f-block, they are also known as the 5f elements. Debate over semantics sometimes leads to the exclusion of actinium and lawrencium, but common usage includes them in the series. The elements are **radioactive** metals with only two (thorium and uranium) found in any significant quantity in nature. The others only appear as radioactive decay products of other elements or are artificially synthesized.

THE ACTINIDES
Part of the f-block of the periodic table, these are the elements found at the bottom of the periodic table.

35

The lanthanides

Like their actinide neighbours on the table, the lanthanides have been the subject of much debate over time, about exactly which **elements** should be included or excluded from their group. They are metals that, like the transition metals, form many coloured **compounds**. Their uses include the manufacture of special glass, complex alloys, powerful magnets and lasers.

WHY IT MATTERS
The lanthanide elements have wide-ranging uses, across multiple industries, but usually in fairly anonymous ways

KEY THINKERS
Carl Gustaf Mosander (1797–1858)
Johan Gadolin (1760–1852)
Carl Axel Arrhenius (1757–1824)

WHAT COMES NEXT
New and improved separation techniques are likely to be the key to the increased utility of the lanthanides

SEE ALSO
The actinides p.62

In 100 words

Found at the very bottom of the **periodic table** in the f-block, the elements with **atomic numbers** from 57 (lanthanum) to 71 (lutetium) make up the lanthanides. Also known as the rare earth elements, these 15 metals are actually not that rare at all. The chemistry of the lanthanides is, in many cases, very similar. This made them difficult to separate from one another. In Greek, *lanthano* means "hidden", and the history of their various discoveries is littered with missteps and false identifications. Their modern chemical applications span a wide range of specialized, niche applications.

Chemical Bonds and Reactions

Within any substance, the particles present are held together by some combination of chemical bonds and forces, acting as "glue" by binding the individual elements or molecules within the compounds. For example, hydrogen atoms (H) and oxygen atoms (O) are connected within water molecules via covalent bonds in a 2:1 ratio, giving the familiar H_2O, while the water molecules themselves are attracted to one another by intermolecular forces. Table salt has sodium ions (Na^+) and chloride ions (Cl^-) held together in a 1:1 ratio by ionic bonds.

When materials are rearranged to form new substances, the chemical bonds within the reactants first need to be broken and then re-formed to make the new compounds. This process is what we call a chemical reaction.

36

Electro-negativity

WHY IT MATTERS
Electronegativity, in large part, governs the chemical reactivity of covalently bonded compounds

KEY THINKER
Linus Pauling (1901–1994)

WHAT COMES NEXT
The ability to predict chemical reactions based upon electronegativity is vital, especially in organic chemistry

SEE ALSO
The nuclear model p.24
Covalent bonds p.70
Reaction mechanisms p.106

A **covalent bond** (see p.70) is a pair of **electrons** held between two **atoms**. Each atom within the **bond** will have its own **electronegativity** value. If the electronegativites are identical, then the electron density will be perfectly evenly spread, and neither end of the bond will develop a positive or negative particle charge. However, when one atom within the bond exhibits an electronegativity that is significantly different to the other end, the electrons within the bond are attracted to the atom with the greater electronegativity. This creates a build up of electron density at one end, and a negative partial charge. This also creates a relatively positive partial charge at the other end. We call this separation of charge a dipole (two poles, one positive, one negative). When this relative imbalance of charge occurs, the bond becomes susceptible to interactions with other positive and negative species, and chemical reactions are likely.

The electronegativity of any given atom depends on multiple factors. The size of the atom and its nuclear charge are two. In general, electronegativity decreases the further one moves away (down and to the right) from the most electronegative atom, fluorine, on the **periodic table**. Since group 18 **elements** tend not to form many covalent bonds, they are usually omitted from electronegativity discussions.

In 100 words

In chemistry, **electrons** are *incredibly* important. It's these tiny subatomic particles that determine the chemical reactivity of **elements** and **compounds**. So, any concepts that deal with electrons are fundamental in chemistry. **Electronegativity** is the measure of how easily an **atom** can attract electrons to itself within a **covalent bond**. Electronegativity is measured on the Pauling scale, named after Linus Pauling who was the first to quantify the property. On the scale, fluorine has the highest value of all the elements, with elements of the bottom of group 1 and group 2, francium and radium, having the smallest values.

37

Ions and ionic bonds

WHY IT MATTERS
Ionic bonding is one of the two most common types of bonding in chemical compounds. The other is covalent bonding

KEY THINKERS
Michael Faraday (1791–1867)
Gilbert Lewis (1875–1946)

WHAT COMES NEXT
Ion technology is vital in modern battery development

SEE ALSO
Electronegativity p.66
Covalent bonds p.70

Without question the most familiar ionic **compound** is sodium chloride – regular table salt. It is made up of a positive sodium **ion** Na^+ (a sodium **atom** that has lost an **electron**), known as a cation, and a chloride ion Cl^- (a chlorine atom that has gained that same electron) known as an anion. Salt exists because of the very strong electrostatic attraction between the oppositely charged sodium and chloride ions.

Ionic bonding is usually observed between metals and nonmetals – i.e., **elements** that tend to be widely separated on the **periodic table**. Pairs of elements such as those tend to have large differences in **electronegativity** (see p.66), meaning that not only does one pull electrons toward itself more strongly than the other, but sometimes the pull is strong enough to completely "steal" the electron, forming two fully oppositely charged ions.

Michael Faraday was the first to propose the term "ion" during his study of electrolysis, and Gilbert Lewis's ideas surrounding the stability of the **noble gases** (see p.60) helped to explain the movement of electrons when atoms (ions) bond ionically. The strong ionic bond, and it being made of positive and negative species, has consequences for the properties of compounds that contain it. They tend to be solids at room temperature, with higher melting and boiling points, and they readily dissolve in water to become good conductors of electricity.

In 100 words

Atoms are electrically neutral particles. This means that they possess an equal number of positively charged **protons** and negatively charged **electrons** (**neutrons** have no charge, and therefore don't contribute to the overall charge of the atom). However, atoms don't always stay in this electrically neutral state. When one atom loses negative electrons to become positive, and passes that negative electron to another atom that then itself becomes negative, they lose their neutral nature. These charged species are called **ions**, and it is the attractions between oppositely charged ions that create what we call **ionic bonds**.

SODIUM CHLORIDE
To form this ionic compound, a sodium atom loses an electron to a chlorine atom, creating a positive sodium ion and a negative chloride ion.

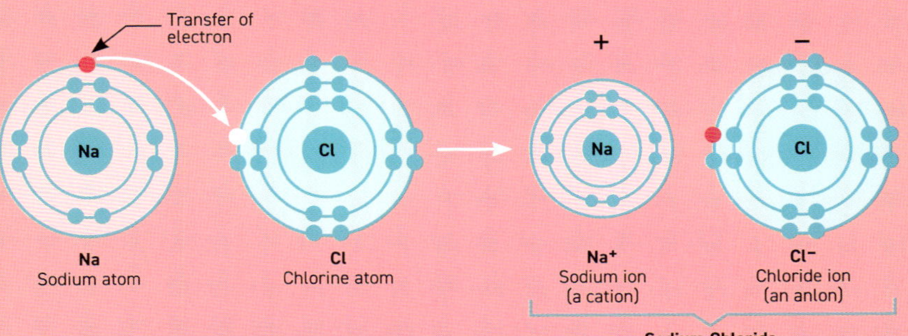

38

Covalent bonds

Where the number of shared pairs between **atoms** is two or three, the **bond** is described as being a double or triple **covalent bond**. As the number of covalent bonds increases, so does the strength of the bond. With increasing strength, bonds become shorter in length. The polarity of bonds, based upon differences in **electronegativity** (see p.66), is an important factor in determining the reactivity of the covalent bond. Gilbert Lewis introduced the idea of **electron** pair sharing in 1916, and Irving Langmuir coined the term "covalence" to describe the shared pairs in 1919.

WHY IT MATTERS
Covalent bonding is the fundamental concept that describes the bonding between nonmetal atoms

KEY THINKERS
Gilbert Lewis (1875–1946)
Irving Langmuir (1881–1957)

WHAT COMES NEXT
Once the bonds between atoms are understood, the chemical reactions that result from the breaking and making of these bonds can be studied

SEE ALSO
Electronegativity p.66
Ions and ionic bonds p.68
Reaction mechanisms p.106

In 100 words

A **covalent bond** is formed between two **atoms** by the sharing of a pair of **electrons**. When the bond is created, there is a balance of electrostatic attraction and repulsion between the positive **nucleus** of each atom and their negative electrons. In many cases, the formation of covalent bonds leads to the attainment of full s and p valence subshells of atoms, otherwise known as "full octets". The octet mimics the outer electronic configuration of the **noble gases**, which are known to be relatively unreactive, meaning that the formation of the covalent bond is motivated by an increase in stability.

HYDROGEN CHLORIDE
In this example, a hydrogen atom and a chlorine atom share a pair of electrons, which increases their stability.

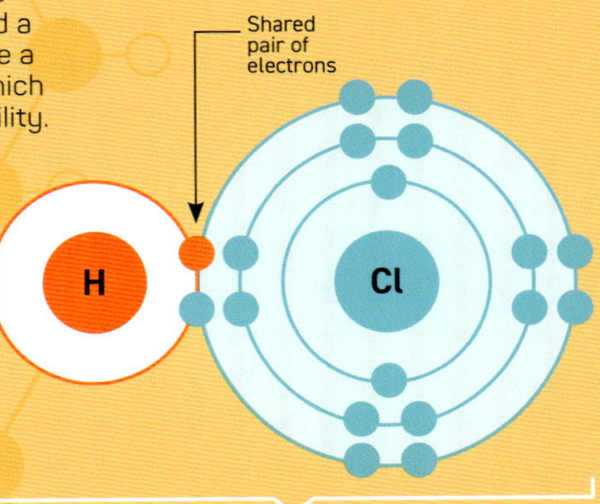

39

Metallic bonds

WHY IT MATTERS
Depending on exactly how they are counted, around 80 per cent of elements are metals, so understanding how they are held together is very important

KEY THINKER
Paul Drude (1863–1906)

WHAT COMES NEXT
The supply of many metals is coming under increasing pressure as Earth's natural resources dwindle in the face of increasing demand from modern technologies

SEE ALSO
Ions and ionic bonds p.68
Covalent bonds p.70

Like any type of chemical bonding, **metallic bonding** must be consistent with the observed properties of any materials that contain it. So, in the case of metals that are malleable, ductile and good conductors of heat and electricity, the ability of the **electrons** to move within the structure while maintaining the bonds, the close-packed array of **ions** and the mobility of the electrons, all contribute to properties that we associate with typical metals.

The discovery of metallic bonding has its roots in the 1900 paper *The Electronic Theory of Metals (Zur Elektronentheorie der Metalle)* by German physicist Paul Drude. In it, he suggested that positive "cores" were surrounded by mobile electrons. This idea has since become known as the "sea of electrons" **model**.

In 100 words

Metallic bonding is the theory that explains how the metal **elements** are held together. It describes an array of closely packed **atoms**, each one releasing its valence (outer) **electrons** into a mobile "sea" of negative charge. Here, the electrostatic attraction between the electrons and the positively charged **ions**, which were left behind when the electrons migrated, acts like a glue to hold the metal together. Both the size and charge of the metal ions (which is to say, the number of electrons that each atom releases) influence the properties of the particular metal, such as its melting point, conductivity and strength.

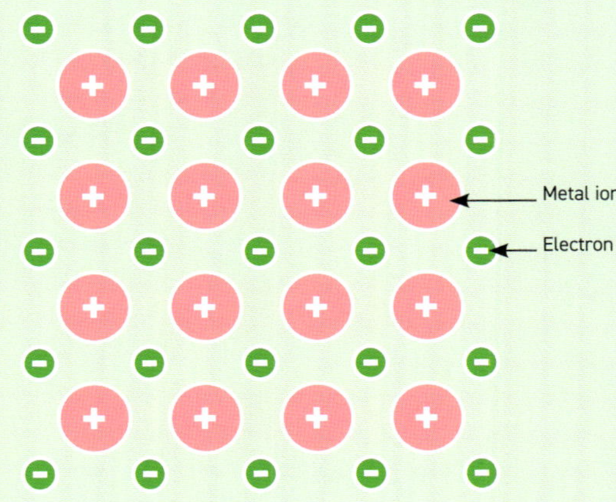

METALLIC BONDING
Metal elements are held together by the electrostatic attraction between positive metal ions in a sea of negative electrons.

40 Molecules

It's hard to pinpoint the moment at which our modern understand of the **molecule** came into being. However, several instances stand out in the history of chemistry. The 17th-century Dutch natural philosopher Isaac Beeckman toyed with the idea of discrete, unconnected particles based upon the ancient Greek ideas of the **four classical elements**. Pierre Gassendi visited Beeckman in 1629 and several years later was the first to use the word *molecule*. In his historic work *The Sceptical Chymist*, Robert Boyle (see p.11) used the phraseology "clusters of corpuscles", which may reasonably be interpreted as having a significant connection to the modern definition of the word *molecule*.

WHY IT MATTERS
An understanding of the particles and entities that make up matter was – and remains – fundamental to the understanding of chemistry as a whole

KEY THINKERS
Pierre Gassendi (1592–1655)
Robert Boyle (1627–1691)

WHAT COMES NEXT
Many unique properties of molecules depend upon the interactions between them, i.e., their intermolecular forces

SEE ALSO
Covalent bonds p.70
Intermolecular forces p.76

In 100 words

The strict definition from IUPAC (see p.32) of a molecule is *an electrically neutral entity consisting of more than one* ***atom***. In most practical instances, this really means a small group of atoms, **covalently bonded** together (see p.70). When the atoms in a molecule are all of the same type, the molecule may be described as **homonuclear** – for example, oxygen (O_2). For more complicated molecules that consist of different types of atoms, we may use the term **heteronuclear** – for example, glucose, ($C_6H_{12}O_6$). Molecules may also be described as **diatomic** (made of only two atoms) or **polyatomic** (consisting of many atoms).

In 100 words

When a chemist observes a reaction on the macro scale, they need a way to describe what is happening at the invisible, atomic level in a concise and accurate way. To do that, they use **chemical equations**. An equation shows the reactants (starting materials) and the products (ending materials) represented by chemical formulae (a shorthand way of describing each substance) with an arrow to show the direction of the reaction. Chemical reactions must show the same number and the same types of **atoms** on each side of the equation so as to adhere to the principle of conservation of mass.

41

WHY IT MATTERS
Equations act as a shorthand method of illustrating what is happening at the otherwise invisible, atomic and molecular level

KEY THINKER
Antoine Lavoisier (1743–1794)

WHAT COMES NEXT
Once equations are written and balanced, the concept of chemical stoichiometry (see p.78) can be applied to determine the amounts of each substance involved in the reaction

SEE ALSO
The conservation of mass p.16
The mole p.78

Chemical equations

Chemical equations not only tell us about the nature of the starting materials and what they turn into, they also offer a way to quantify what is happening. Using chemical equations and the concept of the **mole** (see p.78), we can make predictions about the amount of reactants required and the amounts of products that will be produced. This is crucial in the commercial world of the chemical industry, where chemicals are made for a profit. In the chemical equation shown below, the numbers that precede each formula tell us the number of reacting and product moles, i.e., four moles of ammonia (NH_3) react with five moles of oxygen gas (O_2) to produce four moles of nitrogen monoxide (NO) and six moles of water (H_2O).

$$4NH_3 + 5O_2 \Rightarrow 4NO + 6H_2O$$

42

Intermolecular forces

WHY IT MATTERS
The behaviour and properties of molecules largely depend on their interactions with other molecules. These interactions are called intermolecular forces

KEY THINKER
Fritz London (1900–1954)

WHAT COMES NEXT
Intermolecular interactions are being studied in new areas such as drug design and nanotechnology

SEE ALSO
Electronegativity p.66
Molecules p.74

As **electrons** shift inside **atoms** and **molecules**, specific areas within them can become partially electrically charged; slightly negative in places where the electron density builds up, and slightly positive in places from where the electrons have moved away. All intermolecular forces are based upon the idea that the oppositely charged areas of each molecule will develop an electrostatic attraction for the opposite charges found in another molecule, thus causing them to interact. The opposite charges can be created by differences in **electronegativity** (see p.66), leading to both dipole-dipole attractions and hydrogen bonding, or by the temporary displacement of electron clouds in dispersion forces. These dispersion forces are often called "London" dispersion forces after the German American physicist Fritz London who developed the theory.

In 100 words

Once a **molecule** (see p.74) has been formed by atoms **covalently bonding** (see p.70) to one another, they will have formed discrete entities. These entities (molecules) are attracted to one another through what we call intermolecular forces, or IMFs. These attractions are crucial to an understanding of chemistry since their nature determines many bulk properties, such as melting point, boiling point, vapour pressure, viscosity, volatility and how different molecules interact with one another. Many subtly different types of IMF exist, including hydrogen bonds, dipole-dipole interactions and London dispersion forces, and they account for many of the properties of molecular substances.

43

The mole

WHY IT MATTERS
The mole gives chemists a way to quantitatively link the "invisible" atomic scale to the observable macro scale

KEY THINKER
Amedeo Avogadro (1776–1856)

WHAT COMES NEXT
The mole allows chemists to apply the concept of stoichiometry, which in turn allows prediction of the quantities needed for and produced in chemical reactions

SEE ALSO
The conservation of mass p.16
Element symbols p.44
Chemical equations p.75

When a **chemical equation** is balanced to comply with the conservation of mass (see p.16), there are other consequences to the act of balancing. The numbers that appear in front of each chemical formulae are known as the **stoichiometric coefficients**. These coefficients have great meaning. They indicate the ratio of the **moles** of each species in the reaction. For example, the equation below shows that 1 mole (the "1" is implied and usually not explicitly written) of nitrogen gas (N_2), reacts with 3 moles of hydrogen gas (H_2) to produce 2 moles of ammonia (NH_3).

$N_2 + 3H_2 \leftrightharpoons 2NH_3$

The **periodic table** lists the molar masses of each **element** – i.e., the mass in grams of 1 mole of **atoms** of that particular element. The mole is the unit that can be used to calculate the masses of solids or the volumes of gases in chemical reactions. For example, in the equation above, we know that 28g of N_2 would react with 6g of H_2 to produce 34g of NH_3. Note that mass is conserved.

Avogadro's number can be used to convert between the number of moles and a specific number of particles, using the knowledge that 1 mole of any substance contains 6.022×10^{23} particles.

In 100 words

Avogadro's **mole** concept, named after Italian scientist Amedeo Avogadro, is used to help in the counting of tiny **atoms**. Atoms and **molecules** are so miniscule that, in order to work with macroscopic quantities of them, we need to consider extraordinarily large quantities. The unit known as the *mole* is that large quantity. It is defined as the amount of substance containing the same number of particles as there are atoms in 12 grams of carbon-12. This number, known as Avogadro's number or Avogadro's constant, is equal to 6.022×10^{23}. The mole is one of the standard **SI units**.

44

States of matter

WHY IT MATTERS
Matter matters because it is the stuff that makes up absolutely everything around us!

KEY THINKERS
David Thouless (1934–2019)
Duncan Haldane (1951–)
Michael Kosterlitz (1943–)

WHAT COMES NEXT
Research into new "states", such as superfluids, may have future applications in electronics, medicine or as yet unknown fields

SEE ALSO
The solid state p.82
The liquid state p.84
The gaseous state p.85
Kinetic molecular theory (KMT) p.86

The transitions between states have different names. Solid to liquid is known as melting, and liquid to gas is evaporation, etc., and these phase changes are collectively called physical changes. Physical changes don't create any new substance; they only involve the same substance moving from one state of matter to another. Physical changes are distinct from chemical changes, which actually alter the fundamental chemical make-up of the substance. For example, in a chemical change, water (H_2O) can be broken apart by electricity to produce hydrogen and oxygen gases, i.e., two new substances. Contrast that with water freezing, where liquid H_2O simply becomes solid H_2O, without the creation of any new substance.

Plasma is a high-energy state made up of **ions** (charged particles, see p.68). Plasmas are able to conduct electricity and can respond to magnetic fields. Examples of plasma include lightning, the gas inside "neon" signs, and the Sun.

These ideas about states of matter have been around since the beginning of human time in one way or another, but some of the most modern thinking surrounds the investigation of exotic states, such as superconductors, superfluids or thin magnetic films. In 2016, the Nobel Prize in Physics was awarded to Thouless, Haldane and Kosterlitz for their work in this field.

In 100 words

Generations of school children have been taught about the three states (phases) of matter: solids, liquids and gases. Occasionally a fourth state is included – plasma. The simple idea of substances being made of particles, which have varying energies and varying movement, works incredibly well at predicting the properties of matter. As the closely packed, relatively low energy particles of a solid are heated, they gain energy. When doing so, they start to break the attractions between one another and form a liquid. Further heating gives the particles even more energy, eventually overcoming the forces in the liquid and forming a gas.

STATES OF MATTER
As a substance is heated, the particles gain energy, breaking the attractions between them and causing a change of state.

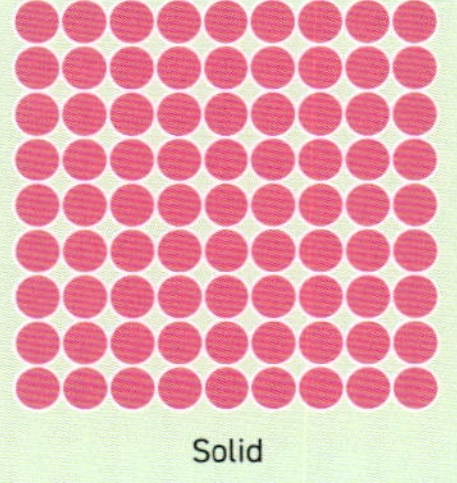

Increasing temperature

45

The solid state

WHY IT MATTERS
Solids are one of the fundamental states of matter. Understanding the particles that make up any given solid and the bonding between them tells us about the solid's properties

KEY THINKERS
Charles-Augustin de Coulomb (1736–1806)
William Lawrence Bragg (1890–1971)

WHAT COMES NEXT
The continued use of solid crystals in laser technology

SEE ALSO
States of matter p.80
The liquid state p.84
The gaseous state p.85

There are many types of solid. Their classification is dependent on their particle composition and the attractions between those particles. For example, you are probably very familiar with table salt (sodium chloride). It is a type of ionic solid where strong attractions between oppositely charged **ions** hold the **crystalline** structure together. This is very different to a solid like glass, which is an amorphous solid, and to a solid like ice, which is made up of **molecules** of H_2O that are attracted to one another much less strongly. The strength of the attractive forces makes a difference to the properties of a solid, with the stronger **ionic bonds** (see p.68) of salt meaning that it has a much higher melting point (801°C) than ice (0°C), which is held together with much weaker intermolecular forces (see p.76). In turn, metals and network solids each have different types of bonding and therefore have different properties when compared to other solids.

The manner in which ions are attracted to one another in ionic solids is governed by Coulomb's law, named for French physicist Charles-Augustin de Coulomb. Distilled to its simplest form, Coulomb's law states that the attraction of two oppositely charged particles depends upon the magnitude of their charges and the distance between them. William Lawrence Bragg won the 1915 Physics Nobel Prize (interestingly along with his father) for his work in determining the structure of crystalline solids with X-rays.

In 100 words

Solids are usually described as being made of particles that have relatively little energy and that vibrate around fixed positions. That is true of **crystalline** solids that have very regular arrangements. Amorphous solids are less ordered in terms of their structures, but the energy and relative movement of their particles is similar to crystalline solids. The energy of the particles in solids is important since it determines that they cannot overcome the attractions they have for one another, and therefore, that the solid doesn't become a liquid (see p.84) in which the particles move around one another freely.

AN IONIC SOLID
Ionic solids are made of oppositely charged ions that are attracted to one another.

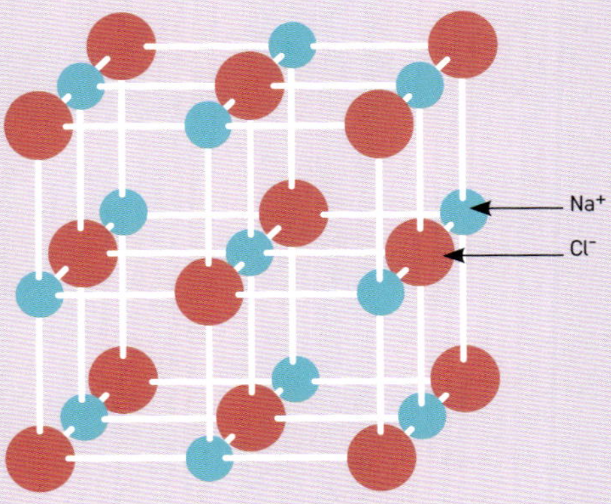

46

The liquid state

Like solids (but unlike gases), liquids contain particles that are very close together, so they are mostly incompressible. Because of their extreme fluidity, the movement of liquids, known as fluid dynamics, is a complex topic. Liquid's fluidity means that attractions are constantly being broken and made between particles, resulting in an extremely "dynamic" or "fluid" substance. Bernoulli's principle relates to fluid dynamics and states that pressure decreases when the flow speed increases. Bernoulli's principle applies to air as well, and it is the basis for the physics behind aircraft wings and flight. As air speeds over the curved upper surface of an aircraft wing, a lower pressure is created on the top surface than on the flatter, bottom surface of the wing, resulting in upward lift.

WHY IT MATTERS
Approximately 71 per cent of the planet's surface is covered by the most ubiquitous liquid – water

KEY THINKER
Daniel Bernoulli (1700–1782)

WHAT COMES NEXT
Liquid metals have many potential applications in medicine and electronics, such as in repairing nerve connections and in thermostats and switches

SEE ALSO
State of matter p.80
The solid state p.82
The gaseous state p.85

In 100 words

Liquids represent an intermediate state of matter between the very ordered and low-energy solid state (see p.82) and the highest-energy, highly disordered state of a gas (see p.85). Liquids have a definite volume but no definite shape; rather, they will take on the shape of their container. As in solids, the particles in a liquid are attracted to one another, but their higher energies mean that they can move past one another freely, resulting in their familiar "flowing" property. If, and when, their energy overcomes the attractive forces between the particles, the liquid particles break free and become a gas.

In 100 words

Gases can be mystifying. The common ones are colourless, making them "invisible". Many have no odour or taste either, meaning they are largely imperceptible to humans. They also have odd properties that liquids and solids don't possess, such as spreading out to occupy the whole space they exist in and being highly compressible. On the other hand, their properties are governed by a very few recognizable variables that have some remarkably simple mathematical relationships. A gas's pressure, volume, temperature and number of **moles** (the amount of it) are linked by several equations known as the gas laws.

The gaseous state

Dealing with invisible and mostly imperceptible things can take a while. The gas laws were formulated by several brilliant minds over a couple of centuries. First up, in 1662, was Robert Boyle (see p.11), who established the relationship between the pressure and volume of a gas – the larger volume, the lower pressure. In 1787, Jacques Charles investigated the relationship between volume and temperature – the higher temperature, the larger volume. In 1808, Joseph-Louis Gay-Lussac found the relationship of pressure and temperature – the higher temperature, the higher pressure. In 1811, Amedeo Avogadro discovered the relationship between the number of moles of a gas and its volume – the more moles, the greater volume. Finally in 1834, Émile Clapeyron combined everything into the ideal gas equation.

47

WHY IT MATTERS
With a countless number of important elements and compounds existing in the gaseous state, an understanding of gases is crucial to everyday life

KEY THINKERS
Robert Boyle (1627–1691)
Jacques Charles (1746–1823)
Amedeo Avogadro (1776–1856)
Joseph-Louis Gay-Lussac (1778–1850)
Émile Clapeyron (1799–1864)

WHAT COMES NEXT
The study of kinetic molecular theory links together the motion of the particles that make up solids, liquids and gases

SEE ALSO
States of matter p.80
The solid state p.82
The liquid state p.84
Kinetic molecular theory (KMT) p.86

48

Kinetic molecular theory (KMT)

WHY IT MATTERS
The ability to predict the behaviour of gases based on a single set of simple assumptions that matches experimental observations, is a powerful one

KEY THINKERS
James Clerk Maxwell (1831–1879)
Ludwig Boltzmann (1844–1906)

WHAT COMES NEXT
Applying KMT to nanotechnology will be central to understanding the interaction of particles at the molecular level

SEE ALSO
The gaseous state p.85

KMT is consistent with the quantitative predictions of all of the gas laws (see p.85), and, as such, it is an excellent **model** that is backed up by experiment. In addition, it allows for the consideration of nonideal gases (those that fail to meet all of the criteria laid down by KMT) and their behaviour. When one of the assumptions of KMT, such as the size of the particles being negligible or the interparticle interactions being nonexistent, fails to be true, we can account for these, both quantitatively and qualitatively, and such observations confirm the validity of KMT. One of the cornerstones to KMT is the Maxwell-Boltzmann distribution, which relates the distribution of the kinetic energies of particles to temperature.

In 100 words

Kinetic Molecular Theory (KMT) explains the observed behaviour of gases by considering the nature, motion and interaction of the **atoms** or **molecules** that make up any gas. KMT assumes that the particles of a gas are in continuous, random motion; that the volume of the gas particles themselves is negligible compared to the total volume that the gas occupies; that there are no interparticle forces between the atoms or molecules; that any collisions between particles are perfectly elastic (no kinetic energy is lost) and that the average kinetic energy of the particles is directly proportional to the Kelvin temperature.

In 100 words

When one substance (the solute) completely dissolves in another substance (the solvent) producing a homogeneous mixture, a **solution** is formed. Mostly commonly, the solvent is water, in which case the solution is referred to as *aqueous*. A homogeneous mixture has no distinguishable parts, and the solute is completely and evenly distributed throughout the solvent. Solutions have significantly different properties to pure solvents. For example, salt water (an aqueous solution of sodium chloride) has a higher boiling point and a lower freezing point than pure water. Such colligative properties were first studied by Richard Watson at Cambridge University (see p.122).

49

Solutions

Solutions are distinct from liquids. Liquids are substances that exist in the liquid state but don't necessarily consist of more than one substance. For example, one can have pure, liquid metal. Solutions are primarily liquids, but not always. A homogeneous mixture of gases like air is also a "solution", just not one you tend to think of as such. When the particles of a dispersed material become distinct from the medium they are in – i.e., they become heterogeneous – they are known as colloids or suspensions. Colloids typically have particles with sizes up to about 500 nm, and suspensions have larger particles, up to about 1000 nm. Colloids are way more familiar to you than you might imagine. The dispersed phase and the dispersion phase can consist of any state. For example, when a gas is dispersed in a solid, we get a solid foam like that found inside a seat cushion. When it's a gas inside a liquid, we get a foam like whipped cream or shaving cream, and when it's a liquid inside a solid, we get a gel.

WHY IT MATTERS
Foams, emulsions, solid foams, gels and liquid aerosols are all colloids

KEY THINKER
Richard Watson (1737–1816)

WHAT COMES NEXT
Colloids are currently being researched as potential drug delivery agents that can target specific cells in the body

SEE ALSO
The liquid state p.84
Colligative properties p.122

Chroma-tography

WHY IT MATTERS
An incredibly important and versatile analytical technique used to analyse mixtures

KEY THINKER
Mikhail Tsvet (1872–1919)

WHAT COMES NEXT
Since chromatography often uses a lot of resources, including potentially harmful solvents, it is one technology that is being targeted for a greener future

SEE ALSO
Intermolecular forces p.76

Mikhail Tsvet was a botanist rather than a chemist and much of his work at the beginning of the 20th century involved plant pigments. He used chalk (calcium carbonate) as the stationary phase and a mixture of organic solvents as the mobile phase. He observed the separation of the coloured **compounds** found in plants and christened the experimental technique *chromatography*. Since the early days of his work, chromatography has greatly expanded into far more sophisticated areas, such as gas chromatography and high-performance liquid chromatography (HPLC). In organic chemistry, compounds are often clear and colourless, so the final chromatograms need to be elucidated with some kind of special light or other chemical compound.

In 100 words

Chromatography is an analytical technique used extensively in organic chemistry. It utilizes differences in the intermolecular interactions of the components of a mixture with both a stationary and a mobile phase. In one version of the technique, a small, concentrated sample of the material to be analysed is placed onto a stationary phase such as silica gel. A mobile, liquid phase made of a solvent mixture is then allowed to pass through the apparatus. The components of the mixture separate according to their affinity for either the stationary phase (travelling short distances) or the mobile phase (travelling relative long distances).

51

Redox reactions

WHY IT MATTERS
Redox reactions are at the centre of plant and animal life

KEY THINKERS
Antoine Lavoisier (1743–1794)
J J Thomson (1856–1940)

WHAT COMES NEXT
As the demand for battery technology increases, the study of redox reactions will become increasingly important

SEE ALSO
Combustion reactions p.92
Batteries p.124
Electrolytic cells p.128

Lavoisier's study of combustion (see p.16) and Thomson's discovery of the **electron** (see p.23) can be seen as crucial milestones in **redox** chemistry, but redox reactions are all around us and at the centre of many familiar, everyday processes. Rusting is a redox reaction, as is the browning of a cut apple. In fact, almost any reaction that results from the exposure to oxygen in the air is a redox reaction. Some key examples include **photosynthesis** and respiration – the very reactions that give life, combustion, battery technology (see p.124) and electrolysis technology (see p.128). The concept of **oxidation** numbers, or oxidation states, is crucial to redox chemistry. The assignment of an oxidation number, typically an integer between +9 and -5, to a chemical species allows the movement of electrons, and therefore redox, to be tracked. If an oxidation number is seen to decrease during a chemical reaction, then negative electrons must have been gained by that chemical species. Therefore the redox must have reduced, and vice versa.

In 100 words

Any chemical reaction where **electrons** are transferred from one chemical species to another is known as a ***redox*** reaction. The word redox is derived from the two processes that are involved – **reduction** and **oxidation**. In electrochemistry, a reduction reaction is defined as one where electrons are gained, and the process of oxidation is one where electrons are lost. These two processes are reciprocal to one another in as much as the electrons gained in the reduction process are the same electrons lost in the oxidation process. The two individual processes are known as half-reactions, and together they make the full redox process.

52

Combustion reactions

WHY IT MATTERS
The ability to burn fuel to convert energy may be the chemical reaction that started civilization itself

KEY THINKER
Antoine Lavoisier (1743–1794)

WHAT COMES NEXT
As climate change continues to put pressure on the use of fossil fuels, using pure combustion for generating energy will likely become less frequent

SEE ALSO
Phlogiston p.14
The conservation of mass p.16
Redox reactions p.90

The increase in the generation of carbon dioxide since the Second Industrial Revolution in the mid-19th century has had a devastating effect on the climate of Earth. It is now recognized as a huge threat to the planet – capable of destroying our natural resources with potentially disastrous consequences for human existence itself. As **hydrocarbons** can't be burned efficiently without producing a damaging greenhouse gas, burning fossil fuels is now understood to be incredibly dangerous, with both governments and individuals having responsibility for limiting or eliminating their use.

GLOBAL CO_2 EMISSIONS FROM HUMAN ACTIVITY
A graph showing the huge increase in CO_2 emissions from 1850.

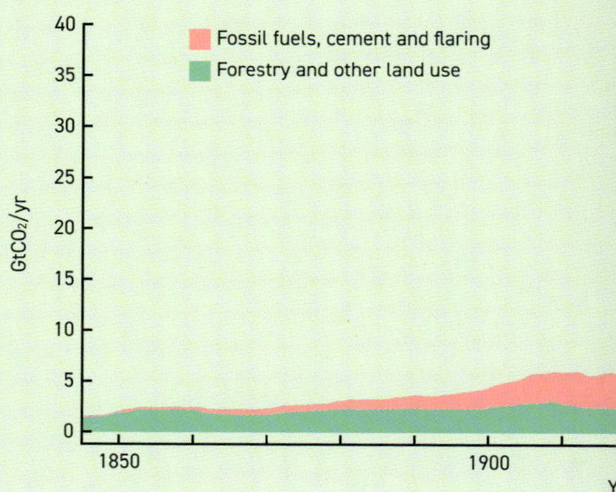

In 100 words

Combustion, more commonly called burning, is a reaction between some kind of fuel and oxygen gas, which produces energy and various products. In the most common cases, the fuel is an organic **compound** such as a **hydrocarbon** (see p.140) or a carbohydrate (see p.160), and the products of the combustion reaction are typically carbon dioxide and water vapour. Compounds such as methane, propane, butane and octane, along with coal, all have extremely **exothermic** enthalpies of combustion. Because they release large amounts of energy as they burn, they have been used for centuries to generate energy.

CUMULATIVE CO_2 EMISSIONS
A comparison of CO_2 emissions between 1750 and 1970 and between 1750 and 2011

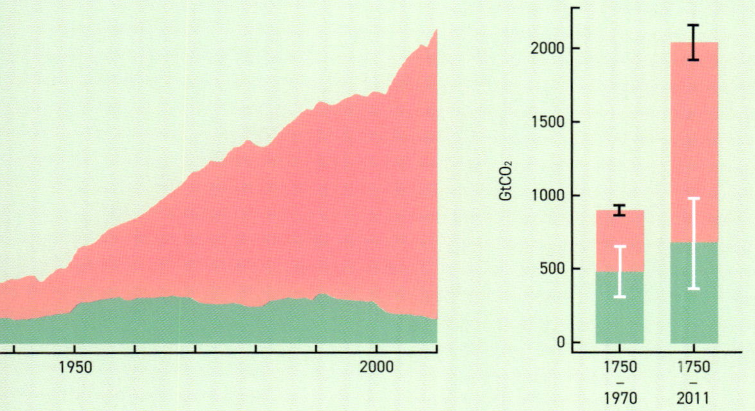

Applying Chemical Reactions

Humankind has been grappling with the application of chemical reactions since the beginning of time. How can we make chemistry work for us in a practical sense? From the harnessing of combustion reactions by making fire to the ultra-sophisticated realm of nuclear chemistry, the study of chemical reactions and energy changes continues to be at the centre of applying the science.

History and chemistry are entangled, with events sometimes driving the chemistry, and chemistry sometimes driving the course of history. Either way, it's hard to deny the enormous influence, both good and bad, that events like the smelting of metals, the development of gunpowder and the production of chemicals, such as fertilizers, dyes, acids and textiles, have had on the world.

53

Chemical energy

In 100 words

Energy is defined as the ability to do work. Chemistry is more specifically concerned with chemical energy – the energy that is either absorbed or released when a chemical reaction takes place. When a reaction occurs, **bonds** are broken in the reactants, and bonds are made in the products. Bond breaking requires the input of energy, and bond making releases energy. The sum of those individual processes determines whether the chemical reaction absorbs or releases energy overall. A reaction that releases energy overall is described as **exothermic**, whereas one that absorbs energy from its surroundings is called **endothermic**.

WHY IT MATTERS
Most chemical reactions bring significant changes in energy

KEY THINKER
James Prescott Joule (1818–1889)

WHAT COMES NEXT
As the planet's energy requirements continue to grow, research into alternative sources of energy dominates many areas of science

SEE ALSO
The first law of thermodynamics p.96
Gibbs free energy p.100

Instant cold packs and instant hot packs are great examples of endothermic and exothermic reactions in action. In each case, the reaction is activated by breaking a vial of a reactant inside the pouch so it mixes with other reactants. An exothermic reaction releases energy to the surroundings and your hands feel the heat. An endothermic reaction absorbs energy from the surroundings (you), and you experience coldness as energy leaves your hands.

English brewer and scientist James Prescott Joule was a pioneering figure in relating energy to mechanical movement. His experiments paved the way for the law of conservation of energy, and the **SI unit** of energy, the Joule, is named after him.

54

The first law of thermodynamics

WHY IT MATTERS
Energy cannot be created – it is only ever converted from one type to another

KEY THINKERS
James Prescott Joule (1818–1889)
Rudolf Clausius (1822–1888)
Germain Henri Hess (1802–1850)

WHAT COMES NEXT
AI, electric vehicles and a greater need for server capacity will continue to put a strain on energy resources

SEE ALSO
The second law of thermodynamics p.97

Thermodynamics as a whole, and in its purest form, is a concept best left in the realm of physics, but in chemistry, one application of the first law of thermodynamics stands tall in terms of its usefulness. Hess's law, conceived by the German chemist Germain Henri Hess in 1840, states that in a chemical reaction, the enthalpy change (essentially the energy change) is independent of the route taken. This means that converting **compound** X directly into compound Z will lead to exactly the same energy change if instead one converts X into compound Y, and then Y compound into compound Z. If the starting and finishing points are the same, any route will lead to the same energy change.

In 100 words

In its simplest form, the first law of thermodynamics states that energy cannot be created or destroyed, rather only converted from one form to another. The amount of energy in the Universe is fixed. For example, the chemical potential energy in petrol can be burned in a combustion reaction (see p.92), where it is converted to thermal energy and kinetic energy. Calculations allow us to see that the energy lost by the petrol is exactly equal to the sum of the various energies that it is converted into, which means energy is conserved.

In 100 words

Stated simply, entropy is a measure of energy dispersal, sometimes described as randomness or disorder. The second law of thermodynamics states that the Universe, when left alone, will always increase in entropy over time. Entropy is a tricky concept to grasp, but it's also an easy thing to observe. A hot drink, left alone in a warm room, will cool down over time. The concentrated energy in the drink flows into the cooler room, until the room and drink each reach the same temperature (known as thermal equilibrium). The energy in the drink has been dispersed, and entropy has increased.

55

The second law of thermo-dynamics

The second law of thermodynamics helps chemists to make predictions. When studying thermodynamics, the Universe is usually split into two parts – first, the system (the tiny piece of the Universe we are studying, usually a chemical reaction) and second, the surroundings (everything else). Knowing that entropy will increase over time, we also know that observing the system decrease in entropy means that the surroundings must increase in entropy in order for the second law to be obeyed and to allow for an overall increase. The only way that this can be achieved is for the surroundings to gain energy, meaning we can predict that the chemical reaction will release energy – that is, be **exothermic** (see p.95).

WHY IT MATTERS
The first and second laws of thermodynamics explain the movement of energy in the Universe

KEY THINKERS
Rudolf Clausius (1822–1888)
William Thomson, Baron Kelvin of Largs (1824–1907)

WHAT COMES NEXT
Some scientists believe that once heat death (maximum dispersal and thermodynamic equilibrium) has been achieved for the whole Universe, it will end

SEE ALSO
The conservation of mass p.16
The first law of thermodynamics p.96

56

The third law of thermo-dynamics

WHY IT MATTERS
The third law gives us a reference point for entropy, defining that all substances in the Universe have positive absolute entropies

KEY THINKER
Walther Nernst (1864–1941)

WHAT COMES NEXT
Research into negative temperatures is a way to examine the properties of states of matter under extreme, controlled conditions

SEE ALSO
The first law of thermodynamics p.96
The second law of thermodynamics p.97

The current lowest recorded temperature achieved is 0.000000000038 K. Negative Kelvin temperatures are a thermodynamic concept that are achievable, but not in the sense that one might think. For example, it doesn't mean that the thermodynamic system is colder than **absolute zero**. It means that the system is hotter than an infinitely positive temperature. If one thinks of temperature as a circle, where the top half of the circle represents a temperature scale of absolute zero to infinity in traditional temperatures, then moving into the bottom half of the circle, off the end of the scale, gives us negative temperatures. These negative temperatures are really just a consequence of the way that we define temperature rather than a redefining of what it means to be extremely cold, so absolute zero is not achievable in terms of an incredibly cold temperature.

In 100 words

In simple terms, the third law of thermodynamics states that the entropy of a pure, perfect **crystal** at a temperature of 0 Kelvin is zero. At this theoretical temperature, all motion of particles within the crystal stops, and there is a single possible configuration – that is, only one microstate is present. Any warming of the crystal, no matter how slight, would lead to the dispersion of energy within it and an increase in its entropy. There are many practical problems with the third law, not least of which is that a temperature of 0 K (**absolute zero**) cannot be achieved.

ENTROPY
A pure, perfect crystal at 0 K is how zero entropy is defined. Any increase in temperature increases its entropy.

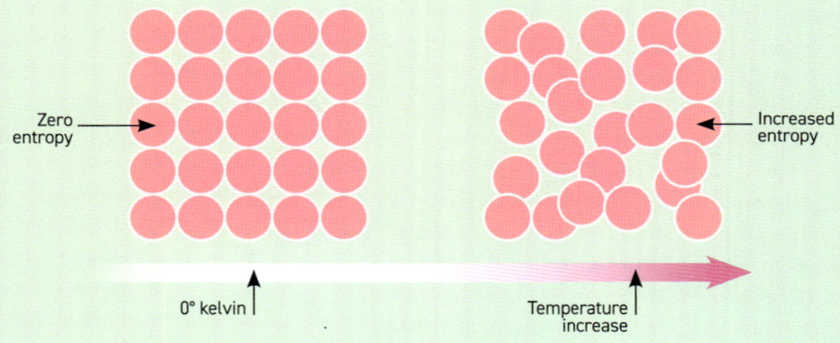

57

Gibbs free energy

WHY IT MATTERS
The likelihood of any chemical reaction can be predicted by Gibbs free energy

KEY THINKER
Josiah Willard Gibbs (1839–1903)

WHAT COMES NEXT
The application of Gibbs free energy to the equilibrium system allows for increased control of chemical reactions

SEE ALSO
First law of thermodynamics p.96
Second law of thermodynamics p.97
Third law of thermodynamics p.98

The Gibbs free energy equation was published in 1873 by the American scientist Josiah Willard Gibbs. The equation allows chemists to consider the entropy change in both parts of the universe (the chemical reaction, which is known as the "system", and everything else in the universe, known as the "surroundings") in order to predict if a chemical reaction is thermodynamically feasible. It considers the entropy of the reaction and the energy change that the surroundings undergo as a result of the reaction. If the reaction releases energy, it is said to be **exothermic** (see p.95) and energy is dispersed into the surroundings, meaning that the entropy of the surroundings increases. **Endothermic** reactions (see p.95) absorb energy from the surroundings, increasing the entropy of the system. When all is said and done, the mathematics of the equation allow us to predict if a chemical reaction is feasible. Any reaction where the Gibbs free energy is shown to be negative is one that is thermodynamically favourable.

In 100 words

The second law of thermodynamics establishes the fact that entropy must increase. So how can a chemical reaction, where the entropy of it decreases, be feasible? The answer is that, in addition to the entropy of the chemical reaction itself (the system), we also have to consider the entropy of the universe outside of the reaction (the surroundings). If a reaction releases lots of energy into the surroundings, then the entropy of the surroundings will increase, which can offset any decrease in entropy of the system, meaning overall entropy increases. Gibbs free energy calculations allow chemists to make these determinations.

58

Chemical kinetics and collision theory

WHY IT MATTERS
Kinetics allows for a greater understanding of the interaction of substances at the molecular level and for greater control of chemical reactions

KEY THINKERS
James Clerk Maxwell (1831–1879)
Ludwig Boltzmann (1844–1906)

WHAT COMES NEXT
More and more advanced catalysts to increase efficiency and speed in commercial processes

SEE ALSO
Catalysts p.104
Reaction mechanisms p.106

Factors that affect aspects of collision theory will, in turn, influence the speed of a reaction. For example, a higher concentration of reactant particles means more collisions and a greater likelihood of successful collisions. An increase in temperature increases the kinetic energy of the reacting particles as predicted by the Maxwell–Boltzmann distribution, which in turn raises more of those collisions above the **activation energy** barrier. Crushing solids to increase their surface area will also result in more collisions and a faster rate. For example, a cube of sugar will take longer to dissolve compared to the same amount of powdered sugar. Finally, any substance that causes the activation energy barrier to be lowered means that more collisions are likely to reach that lower barrier, and once again, the reaction speed will increase. Such substances are called **catalysts** (see p.104).

In 100 words

Some chemical reactions occur extremely slowly, such as rusting. Others are essentially instantaneous, such as explosions. Chemical kinetics is the study of the speed of reactions. Collision theory governs kinetics, and its first criterion is that reacting particles must collide. However, a collision alone is not necessarily sufficient for the reaction to take place. For a collision to result in the formation of products – that is, for it to be successful – it needs two characteristics. First, a certain minimum energy known as the **activation energy**, and second, a certain physical orientation that aligns reactants correctly in three-dimensional space.

ENERGY PROFILE DIAGRAM
Plots the changes in energy of the species in a chemical reaction as the reaction proceeds.

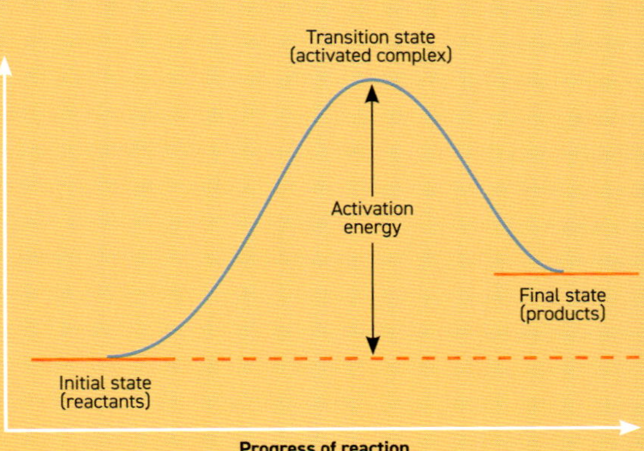

59

Catalysts

WHY IT MATTERS
Many commercial chemical reactions and processes depend on timely execution

KEY THINKERS
Karl Ziegler (1898–1973)
Giulio Natta (1903–1979)

WHAT COMES NEXT
As the alternative energy industry gathers pace, catalysts for producing renewable energy, such as hydrogen production, will increase in importance

SEE ALSO
Chemical kinetics and collision theory p.102

Biological **catalysts** are incredibly important, and many critical functions rely upon them. In biology, it is usually **enzymes** that work as catalysts. For example, pepsin in the human digestive system catalyzes the breakdown of proteins in food, aiding the digestion process. Lactase is another important enzyme in the body. Without it, the breakdown of lactose in the body fails to work correctly, making a person "lactose intolerant" and causing symptoms such as bloating and general gut discomfort.

Outside of the body, examples of catalysts in industry include platinum metal in the catalytic converters of cars and zeolites that catalyze petroleum cracking (see p.138). Ziegler–Natta catalysts were discovered by Karl Ziegler and Giulio Natta in the 1950s. They typically contain transition metals (see p.58) and organometallic **compounds**, and they are used in the production of polymers (see p.154). They are among the most important catalysts used in commercial applications.

In 100 words

A **catalyst** is a substance that speeds up a chemical reaction. Many chemical processes are dependent upon catalysts to make them commercially viable, and some reactions are practically impossible without them. Examples in industry include nickel in the production of margarine and iron in the Haber–Bosch process (see p.112). Catalysts have two characteristics. Firstly, they are not consumed during the process, meaning that they can be recovered at the end of the reaction and reused. Secondly, they work by lowering the **activation energy** (see p.103), meaning that more of the reactant particles possess this lower energy and can go on to react.

CATALYSTS
A catalyst acts by providing an alternative pathway for the reaction that has a lower activation energy.

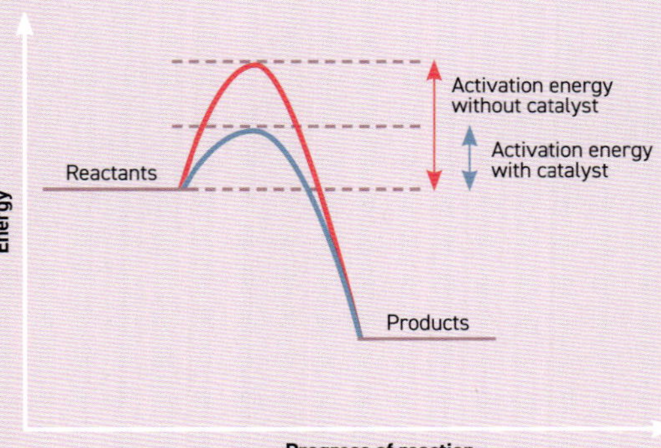

Reaction mechanisms

WHY IT MATTERS
An understanding of the miniscule, detailed specifics of a reaction allow it to be modified and controlled at the atomic level

KEY THINKERS
Arthur Lapworth (1872–1941)
Robert Robinson (1886–1975)
Christopher Kelk Ingold (1893–1970)
William Kermack (1898–1970)

WHAT COMES NEXT
As the field of computational chemistry grows, a deeper understanding of all reaction mechanisms is likely to follow

SEE ALSO
Organic chemistry and vitalism p.136
Computational chemistry p.167

In organic chemistry, the representation of reaction mechanisms is based on the use of "curly arrows". Curly (or curved) arrows are used to indicate the movement of **electrons**. The tail of the arrow shows where the electrons are flowing *from*, and the head shows where they are going *to*. A full-headed arrow indicates the movement of a pair of electrons, and a half-headed arrow shows the movement of a single electron. The use of curly arrows to illustrate mechanisms was proposed by numerous chemists at various points during the 1920s, with Robinson, Ingold, Kermack and Lapworth being some of the prominent players.

In 100 words

Reaction mechanisms describe the detailed, atomic-level interactions of **bond** breaking and making that happen when a reaction takes place. Mechanisms identify each elementary (individual) step that takes place, the sequence of those steps, the precise bond breaking and making and any transition states or intermediates that are formed along the way. Taken together, the elementary steps add up to give the overall chemical reaction. Reaction mechanisms are particularly useful in organic chemistry, where they are used to make predictions about new reactions and bring an understanding that can help to control and enhance already known reactions.

REACTION MECHANISMS
Reaction mechanisms use "curly arrows" to denote the movement of electrons during chemical reactions.

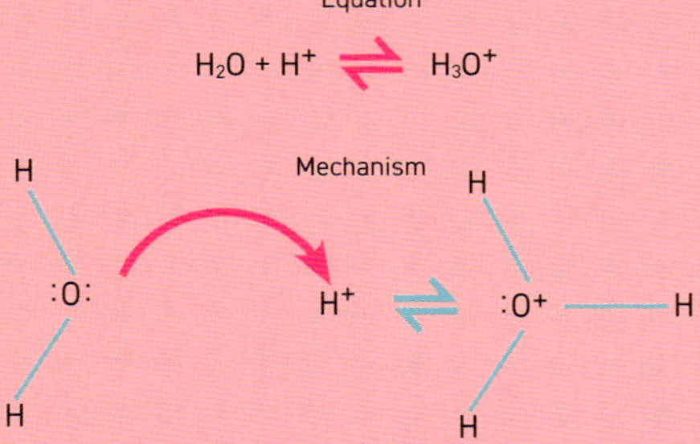

61

Dynamic equilibrium

WHY IT MATTERS
The manipulation of equilibrium reactions is crucial in industry because by controlling conditions we can favour the creation of products

KEY THINKERS
Claude Louis Berthollet (1748–1822)
Henry-Louis Le Chatelier (1850–1936)

WHAT COMES NEXT
Controlling equilibrium mixtures by varying the conditions and application of Le Chatelier's principle (see p.110)

SEE ALSO
Le Chatelier's principle p.110
The Haber-Bosch process p.112

Once an equilibrium has been achieved, there are several things to consider. The first is that "equilibrium" doesn't mean that there are an "equal" number of reactants and products; rather, it refers to an equal rate of forward and backward reactions, and so the status quo of reactant and product concentrations is maintained. When the equilibrium is achieved, it's entirely possible that there is a large conversion of reactants to products, or vice versa. To know the relative amounts of reactants and products that have been formed at the equilibrium position, we use something called the equilibrium constant. It is the ratio of one aspect of the products (usually concentration or pressure) to the same aspect of the reactants, expressed as a numerator (products) and denominator (reactants). The equilibrium constant is given the symbol K, and large values much greater than 1 suggest large amounts of products at equilibrium, and vice versa.

Equilibria occur in nature all the time, and one crucial one happens in the human body. An equilibrium is set up in blood to keep the **pH** (see p.118) between approximately 7.3 and 7.5, so it never gets too low or too high and the status quo is maintained. In industry, the production of ammonia in the Haber-Bosch process (see p.112) is an incredibly important application of an equilibrium system, where the reactants nitrogen and hydrogen react together to form ammonia gas until an equilibrium is reached.

In 100 words

Some chemical reactions just move in one direction, by only converting the reactants to the products. Other reactions have a tendency to also convert products back to reactants. When the forward reaction and the backward reaction occur at the same rate (see p.102), we say that an equilibrium is achieved. At the macro level, it appears as though nothing is happening since reactants are used up in the forward reaction at exactly the same rate at which they are replaced in the backward, but at the **molecular** level, the reaction is still happening in both directions. This is called a dynamic equilibrium.

62

Le Chatelier's principle

WHY IT MATTERS
This principle allows chemists to predict how equilibrium systems respond to changes, and it is especially important in industry to maximize yields and reduce costs

KEY THINKER
Henry-Louis Le Chatelier (1850–1936)

WHAT COMES NEXT
Once we know how far a reaction will go, it is helpful to know how fast it will go (see p.108)

SEE ALSO
Dynamic equilibrium p.108

The principle was formulated by French chemist Henry-Louis Le Chatelier in 1884. In 1888, he revised his initial declaration to simplify it. It can be applied to changes in concentration, pressure and temperature, but in each case it is always about restoring an equilibrium position that has somehow been disturbed. For gaseous reactions, increasing pressure will shift the system towards the side of the reaction with fewer gas **molecules**, thus decreasing pressure and opposing the change. Temperature changes will shift the system according to the **endothermic** or **exothermic** nature of the system. An increase in temperature will shift a reaction in the endothermic direction, and vice versa.

In 100 words

Le Chatelier's principle is a conceptual idea that is applied to chemical equilibrium systems. The principle states that when a chemical system, which has achieved equilibrium, is subjected to an external stress of some kind, the system will shift in such a way as to oppose that change, and in the process re-establish the equilibrium position. For example, if the external stress happens to increase the concentration of the reactants in the system, then the system will shift to the product side, converting those extra reactants into products and so counteracting the stress and restoring equilibrium.

In 100 words

Aqua regia (*royal water* in Latin) has a special place in the history of chemistry, in part because it remains one of the few liquids capable of dissolving gold. As such, it was a substance revered by the alchemists since they saw it as being linked to their quest for the philosopher's stone (see p.10). With its existence known since the 13th century, aqua regia is not chemically complicated, being nothing more than a mixture of 25 per cent nitric **acid** and 75 per cent hydrochloric acid. The caustic mixture will dissolve platinum and gold, and that property was put to good use during World War II.

63

Aqua regia

When the Nazis marched into Denmark in 1940, one of their goals was to repatriate (or loot) any gold that they could find. Two German Nobel Prize winners in physics, Max von Laue and James Franck, had sent their solid gold winners medals to The Neils Bohr Institute in Copenhagen. Hungarian chemist George de Hevesy, who was working at the institute at the time, had the bright idea of dissolving the medals in aqua regia. Not only did he succeed and simply leave the resulting innocuous-looking pale orange liquid on a shelf in his laboratory, he managed to retrieve the liquid years later, precipitate out the gold and send it to the Nobel institute in Sweden, where it was recast into medals and once again presented to the Nobel winners.

WHY IT MATTERS
One of the most enduring substances in chemical history, dating from the era of alchemy to the present day

KEY THINKER
George de Hevesy (1885–1966)

WHAT COMES NEXT
Aqua regia's role in recovering precious metals like gold from used electronics has become crucial in environmental and green chemistry

SEE ALSO
Alchemy p.10
Acids, bases and neutralization p.114

64

The Haber-Bosch process

WHY IT MATTERS
The more efficient, industrial-scale production of ammonia for making fertilizers revolutionized agriculture and world food production

KEY THINKERS
Fritz Haber (1868–1934)
Carl Bosch (1874–1940)

WHAT COMES NEXT
With the production of ever more efficient catalysts, the use of lower temperatures and pressures in the process will be achievable

SEE ALSO
Dynamic equilibrium p.108
Catalysts p.104

The Haber-Bosch process is often used as a case study for the chemical concept of equilibrium (see p.108) and catalysis (see p.104). Equilibrium theory determines that the reaction would yield the greatest amounts of ammonia when performed at incredibly high pressures and very low temperatures, but these conditions are rarely used. Creating high pressure costs money, and running at low temperature slows down the reaction, neither of which are commercially desirable. To make the process viable, compromise conditions are used, along with equilibrium considerations and an important **catalyst**, iron.

Haber's fame goes well beyond his genius in developing this process, for which he won the 1918 Nobel Prize in Chemistry, and stretches to infamy. He is considered by many as the "father of chemical warfare", having developed chlorine and mustard gas as weapons of war in World War I, which was used against the Allies in 1915 at Ypres in Belgium with devastating effect. This was decades before the Nazis seized power in Germany, and Haber is sometimes (wrongly) accused of being a Nazi; he wasn't, he was Jewish, and indeed was forced to flee the country when the Nazis came to power. However, one terrible lingering truth has continued to haunt his legacy. Prior to the Holocaust, Haber had worked on the production of Zyklon B – a gas that was to be used as a pesticide. The tragic end to the story is that the production of Zyklon B morphed into a gas of ultimate destruction used in concentration camps to exterminate Jews.

The role of Carl Bosch – also a Nobel Prize winner in Chemistry – in the Haber–Bosch process is often overlooked, but his contribution was vital. He worked on more efficient sourcing of raw materials, the development of the catalyst and scaling the process from a laboratory reaction to an industrial one.

In 100 words

The Haber–Bosch process was developed by Fritz Haber and Carl Bosch in the early 20th century. The industrial process makes ammonia (NH_3) – a key component of fertilizers – from nitrogen (N) and hydrogen (H) gas. Its mass production revolutionized agriculture and the large-scale production of food across the globe. Ammonia also had a significant impact on World War I. Prior to the war, Germany had been reliant upon imported nitrates for the manufacture of explosives and fertilizers. The domestic production of ammonia allowed the Germans to extend the war and continue to produce large quantities of wartime explosives.

Acids, bases and neutralization

WHY IT MATTERS
Two of the most important general groups of chemicals, acids and bases are essential in our everyday life

KEY THINKERS
Johannes Nicolaus Brønsted (1879–1947)
Thomas Martin Lowry (1874–1936)
Svante Arrhenius (1859–1927)
Gilbert Lewis (1875–1946)

WHAT COMES NEXT
The interdependent relationship between acids and bases leads us to water

SEE ALSO
Water p.116
The pH scale p.118

Some **acids** are incredibly corrosive, being able to destroy metals or human flesh on contact, while others are safe to even eat. Two factors determine an acid's treacherousness: its strength and its concentration. An **acid's strength** is determined by its ability to donate hydrogen **ions**. Acids whose **molecules** donate 100 per cent of their hydrogens are said to be strong, while those that donate much smaller percentages (often less than 5 per cent) are said to be weak. An **acid's concentration** is a measure of how diluted it is. Large amounts of acid dissolved in small amounts of water are said to have high concentrations, and vice versa. It's possible to have a weak, concentrated acid, and a strong, dilute acid.

The role of water in acid-base chemistry is an essential one. Acids in the form of hydrogen ions (H^+) and **bases** in the form of hydroxide ions (OH^-) may come together in a chemical reaction known as a neutralization to produce water (H_2O). This makes the description of acids and bases as the chemical opposites of one another a logical one. The relationship between acids and bases can also be seen via what are known as acid-base conjugate pairs. In an acid-base reaction, it is possible to identify such a pair of substances whose formulae differ only by H^+, for example NH_4^+ (the acid) and NH_3 (the base).

In 100 words

An **acid** is a substance that releases hydrogen **ions** (H+) and has a **pH** level of less than 7. It is commonly defined in one of three ways: Brønsted–Lowry acid, Arrhenius acid or Lewis acid. A Brønsted–Lowry acid, donated by an hydrogen **ion**, is any substance that can transfer a **proton** to any other **compound**. An Arrhenius acid produces hydronium (H_3O^+) ions in **solution**, and a Lewis acid is a substance that accepts a pair of **electrons**. **Bases** are the chemical opposites of acids, with a pH level greater than 7. They accept hydrogen ions, produce hydroxide ions (OH-) in solution and can act as electron pair donors.

STRENGTH OF ACIDS
Strong acids dissociate into their ions completely, while weak acids only partially dissociate. (Water molecules are omitted from the diagrams.)

In water, the strong acid fully dissociates into Hydrogen (H^+) and Chlorine (Cl^-) ion components.

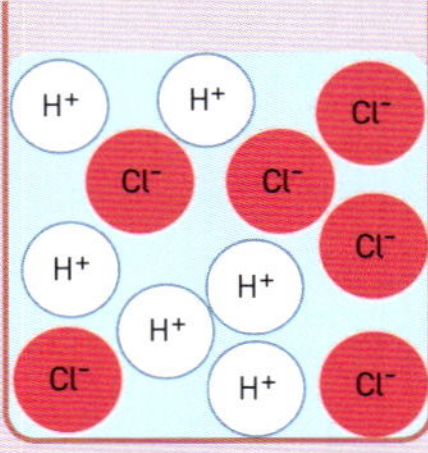

Hydrochloric acid (strong)

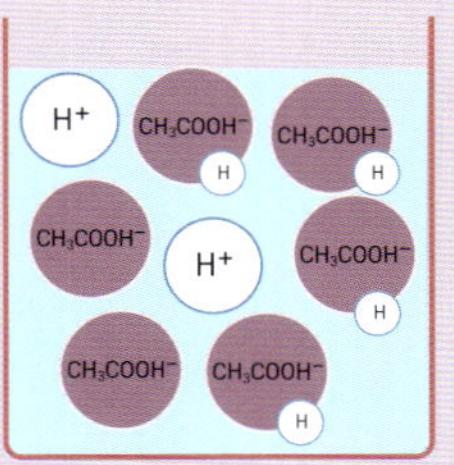

Ethanoic acid (weak)

In water, the weak acid only partially dissociates.

Water

WHY IT MATTERS
Essential to life – without it there is none

KEY THINKERS
Adriaan Paets van Troostwijk (1752–1837)
Jan Rudolph Deiman (1743–1808)

WHAT COMES NEXT
Keeping our water clean is already one of the greatest challenges for chemistry

SEE ALSO
Acids, bases and neutralization p.114
Solutions p.87

Water has many properties that surprise. For example, by scientific prediction, it should have a boiling point of close to −100°C (−212°F), approximately 200°C (392°F) lower than the normal boiling point that we are familiar with of 100°C (212°F). It is just as well it defies prediction, because otherwise all of the water on Earth would be a gas rather than a liquid. That anomalous boiling point is due to an interaction between the water **molecules**, known as hydrogen bonding, and it is also responsible for the fact that solid water is less dense than liquid, and that's why ice floats in water. Water has an extraordinary ability to dissolve many substances, and it has been called the universal solvent as a result. This is less true when it comes to many organic **compounds**, but even those have vast numbers of compounds that will form **solutions** in water. All of these properties and others can be explained by the polarity of water molecules.

In 100 words

Ubiquitous, simultaneously simple and complex, water (H_2O) is probably one of the most universally recognized chemical **compounds** found on Earth. Essential for life, water covers close to 75 per cent of the surface of our planet. It exists as a V-shaped **molecule** with two hydrogen **atoms**, each **bonded** to a single, central oxygen atom. The shape and the arrangement of **electrons** within the molecule create tiny positive and negative electrical charges on its surface, and it's these charges that make water a polar molecule. Water's polarity gives its molecules many of their unique properties.

WATER
Water molecules consist of two hydrogen atoms, each bonded to the same oxygen atom.

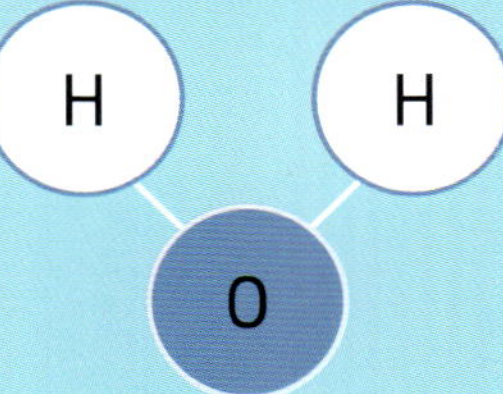

67

The pH scale

WHY IT MATTERS
The pH scale allows for acids and bases to be quantified

KEY THINKER
Søren Sørensen (1868–1939)

WHAT COMES NEXT
With a way to measure acidity or basicity, better management of the environment, organisms and human health could follow

SEE ALSO
Acids, bases and neutralization p.114
Water p.116

While working at the Carlsberg Research Laboratory in Copenhagen, Denmark, Søren Sørensen developed the **pH scale** in 1909. Established in 1875, the laboratory remains a biochemical research centre whose chief purpose has always been to study the science of brewing and fermenting. Sørensen, a Danish chemist, was its director from 1901 to 1938. The pH scale simply converts one measurement (hydrogen **ion** concentration) into another, more manageable number. Pure water has the ability to self-ionize, producing exactly equal amounts of hydrogen ions (H+), which account for acidity, and hydroxide ions (OH-), which account for basicity. At 25°C (77°F), the amounts of each ion are identical, which makes pure water neutral. At other temperatures, the concentrations of the two ions are found to have different but still identical values, meaning that pure water can both be neutral and have pHs other than 7. Universal indicators are chemicals whose colour changes, and so they can be used as a visual aid in determining pH.

In 100 words

The **pH scale** is a scale, usually ranging from 0 to 14, that allows for the classification of a **solution** as being acidic, basic or neutral. A number on the scale below 7 indicates that an **acid** is present; one above 7 is classified as a **base** (the chemical opposite of an acid) and one with a pH equal to 7 is described as being neutral, i.e., neither an acid nor a base. The pH scale is a logarithmic one, where pH = -log [H+], and where [H+] is the concentration of hydrogen **ions** in the solution.

68

Buffers

WHY IT MATTERS
Many important reactions depend on the pH of the environment they occur in, so being able to control it is crucial

KEY THINKERS
Lawrence Henderson (1878–1942)
Karl Hasselbalch (1874–1962)

WHAT COMES NEXT
Buffers can be used to mimic the pH of the human body, thus allowing drug activity to be researched

SEE ALSO
Acids, bases and neutralization p.114
The pH scale p.118

A buffer **solution** has two components mixed together; typically a weak **acid** mixed with its conjugate **base** or a weak base mixed with its conjugate acid. When small amounts of acid or base are added to such a solution, this additional acid or base is absorbed by the base or acid component in the buffer. This means that large changes in **pH** are resisted, and only relatively tiny increases or decreases are observed, maintaining something close to the original pH.

The pH of a buffer solution can be determined using the Henderson-Hasselbalch equation, which was developed by Lawrence Henderson and Karl Hasselbalch in the first decade of the 20th century.

In 100 words

Buffers are special **solutions** that exist in acid-base chemistry. When adding an **acid** or a **base** to a solution that isn't a buffer, the **pH** of the solution will change – sometimes dramatically. For example, adding just a few drops of a strong acid to neutral water can decrease its pH from 7 to less than 1. A similarly dramatic increase can occur when adding a small amount of a strong base to water. Buffers resist such dramatic changes in pH when acids or bases are added. They are crucial in many situations but especially in biological processes.

HENDERSON-HASSELBALCH EQUATION
This equation determines the pH of a buffer solution.

$$pH = pKa + \log \frac{(\text{conjugate base})}{(\text{acid})}$$

Colligative properties

WHY IT MATTERS
It's why we "salt" the roads in winter

KEY THINKER
Jacobus Henricus van 't Hoff (1852–1911)

WHAT COMES NEXT
Beetroot juice has been proposed as an alternative to salt for decking roads, but it too creates environmental concerns

SEE ALSO
Solutions p.87

The application of one colligative property happens all over the world, in any place where it gets cold during the winter. Pure water will freeze at 0°C, but adding a salt, often sodium chloride or calcium chloride, lowers the freezing point – to -10°C or lower. This means that the air temperature must be that much colder before the water will freeze, and any ice that does form on the roads, causing a hazard, will melt more easily. At the other end of the scale, adding salt to water raises its boiling point above 100°C, so food will cook more quickly.

In 100 words

Colligative properties are properties of **solutions** that depend only upon the number of solute particles present, not on their nature. This means that seemingly dissimilar things like salt and sugar can have similar effects upon the properties of solvents (most often water) in which they are dissolved. Colligative properties include boiling point elevation, freezing point depression, vapour pressure lowering and osmotic pressure. With an increasing number of solute particles, the particular property manifests itself more strongly. For example, the boiling point of water is continually raised as greater amounts of salt are added.

In 100 words

Sulphuric **acid** has been an important chemical for centuries, appearing in such industries as fertilizer manufacture, chemical production and water treatment. The ability to produce it on a large scale via an efficient process was revolutionized in 1831 when Peregrine Phillips, a vinegar manufacturer, patented the contact process. This method involves converting sulphur dioxide gas into sulphur trioxide gas by passing it over a metal **catalyst**. Prior to the contact process, sulphuric acid had been made using the lead chamber process. The contact process greatly improved things by being more efficient and for allowing the production of more concentrated acid.

70

The contact process

The contact process replaced John Roebuck's lead chamber method that had held sway for about 100 years prior. Roebuck's use of cheaper lead had replaced fragile glass vessels and allowed for larger amounts of sulphuric acid to be made. The contact process took things a step further. Phillips's original process used a platinum catalyst to convert sulphur dioxide to sulphur trioxide – a key step in the process. But platinum was expensive, and it had a tendency to become poisoned and stop working. Over time, platinum was replaced with a vanadium oxide catalyst (V_2O_5). The speed, and therefore the efficiency and cost effectiveness provided by the catalysts, was key to the contact process becoming the dominant manufacturing technique for the production of sulphuric acid.

WHY IT MATTERS
Sulphuric acid is arguably the single most important chemical in industry, with applications as a vital ingredient and precursor in innumerable processes

KEY THINKERS
John Roebuck (1718–1794)
Peregrine Phillips (1800–1888)

WHAT COMES NEXT
The global market for sulphuric acid is only increasing

SEE ALSO
Catalysts p.104
Acids, bases and neutralization p.114

Batteries

The term "battery" was coined by Benjamin Franklin in 1749, but the first true battery was invented by Alessandro Volta in 1800. His voltaic pile used copper, zinc and brine to produce an electrical current. In 1859, the French physicist Gaston Planté invented the lead-acid battery. It was notable for its ability to be recharged. Since then, the humble battery has come a long way, including evolving into the nickel-cadmium battery and the alkaline battery. In the modern world, the lithium-ion battery reigns supreme, being found in all portable electronics and in electric vehicles. The 2019 Nobel Prize in Chemistry was awarded to Stanley Whittingham, John Goodenough and Akira Yoshino for their work in bringing such batteries to commercial reality over several decades.

WHY IT MATTERS
As the world continues to turn away from fossil fuels, the need for greater battery technology will only increase

KEY THINKER
Alessandro Volta (1745–1827)

WHAT COMES NEXT
The continued development of battery technology around new materials such as lithium iron phosphate and sodium ion cells

SEE ALSO
Redox reactions p.90

In 100 words

Some chemical reactions involve the transfer of **electrons** from one substance to another. Such reactions are known as **redox** reactions (see p.90). It's possible to physically arrange the two substances in such a way that the electrons flowing from one reactant to the other also pass through an external circuit and through a device of some sort. Electrons flowing in the external circuit, and so through the device, is what we call electricity. The physical construction of the apparatus used to convert the chemical energy of the reactants into the electrical energy is call a battery.

In 100 words

Standard **reduction** potentials are measured by using the standard hydrogen electrode, which operates at a temperature of 298 K, with any **solutions** having a molarity of 1.0 mol dm$_{-3}$, and any gases having a pressure of 100 kPa (kilopascal). The voltages of electrochemical cells are calculated based upon standard reduction potentials. Whenever any of those conditions deviate from standard values, a non-standard half reaction is created, and a non-standard electrochemical cell will be constructed. When this happens, we must consider the effect of the new conditions and calculate the new voltage of the battery using the Nernst equation.

72

WHY IT MATTERS
Nernst's equation allows us to calculate the voltage of batteries over a range of conditions, not just standard ones

KEY THINKER
Walther Nernst (1864–1941)

WHAT COMES NEXT
In a post-fossil-fuel world, the need to understand the chemistry of electrical cells is growing exponentially

SEE ALSO
Batteries p.124

The Nernst equation

Not only famous for his eponymous equation, German chemist and physicist Walther Nernst also worked on numerous aspects of thermodynamics. In 1920, he was awarded the Nobel Prize in Chemistry for his work in thermochemistry. He penned what became known as the third law of thermodynamics, which has its origins in the Nernst heat theorem. Nernst worked on the theorem in the early years of the 20th century and came to his conclusions in 1912.

$$E_{cell} = E^0 - \left(\frac{RT}{nF}\right) \ln Q$$

73

Hydrogen fuel cells

WHY IT MATTERS
With the phasing out of fossil fuels, hydrogen fuel cell technology will be central to the development of cleaner energy

KEY THINKER
Sir William Robert Grove (1811–1896)

WHAT COMES NEXT
The development of catalysts to aid the splitting of water to yield the hydrogen needed for the cells

SEE ALSO
Batteries p.124

One of the major disadvantages of using hydrogen gas is the problem associated with getting the gas to the end user. This is especially true of the fuel cells used in cars. The infrastructure surrounding the storage and transporting of gases can be complex and expensive. Fuel cells also require the use of some expensive metals for catalysis, which can be both costly and difficult to acquire.

In 100 words

A hydrogen fuel cell generates electricity via a chemical reaction between hydrogen and oxygen gases to produce water. The process is a **redox** reaction (see p.90), by which **electrons** are transferred from the hydrogen to the oxygen. Hydrogen gas is delivered to the anode, where it releases electrons and forms hydrogen **ions** (H_+). The ions pass through the cell to the cathode, while the electrons flow through an external circuit. At the cathode, the hydrogen ions and electrons combine with oxygen to form water. The major advantage of these fuel cells is that their only by-product is water rather than any harmful emissions.

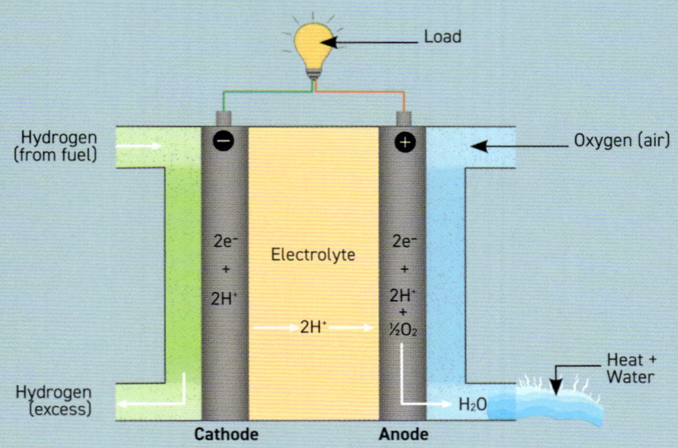

FUEL CELL
In a hydrogen fuel cell, the chemical reaction between hydrogen and oxygen is used to generate electricity and water is a by-product.

74

Electrolytic cells

WHY IT MATTERS
Being able to force a redox reaction to occur gives us control of many industrially important and useful electrochemical processes

KEY THINKERS
Charles Martin Hall (1863–1914)
Paul Héroult (1863–1914)
Humphry Davy (1778–1829)

WHAT COMES NEXT
In the search for renewable energy sources, such as hydrogen gas, breaking water apart via electrolysis is a crucial step

SEE ALSO
Redox reactions p.90
Batteries p.124
Hydrogen fuel cells p.126

The whole process of forcing **redox** reactions to occur by applying an external source of electricity is known as electrolysis. Electrolysis can be used for electroplating metals onto objects and for purifying them via a process called electrorefining. In the realm of chemical production two vital electrolysis processes are the Hall–Héroult cell and the chlor-alkali cell.

In the chlor-alkali cell, brine (salt water) is converted to three useful products: chlorine gas, hydrogen gas and the important industrial **base**, sodium hydroxide. The Hall–Héroult cell is used to extract aluminium metal from its ore. In the mid-1800s, aluminium was considered to be a precious metal since its extraction was so expensive. The Hall–Héroult cell was revolutionary in 1886, when it was developed independently by Charles Hall in the USA and Paul Héroult in France. The electrolytic cell made the large-scale production of aluminium commercially viable thanks to its ability to scale the process and because it produced relatively pure aluminium. Humphry Davy was another important figure in electrolysis. In the early 19th century, he had succeeded in extracting a number of familiar metals, including sodium, potassium and calcium, by using electricity.

In 100 words

If a **redox** reaction (see p.90) is one that doesn't occur spontaneously, then it must be driven by the use of electricity. Such an electrochemical device is known as an electrolytic cell. Compare this situation to a redox reaction that creates electricity from a spontaneous redox reaction. When an external voltage is applied to the electrolytic cell, the chemical species present in the cell, usually water and **ions**, migrate to the oppositely charged electrodes. At the cathode, they are reduced via the gain of **electrons**, and at the anode they are oxidized by losing electrons.

75

Radioactivity

WHY IT MATTERS
The understanding and control of radioactivity is crucial to human well-being for safety and for harnessing its power to work for us

KEY THINKERS
Pierre Curie (1859–1906)
Marie Curie (1867–1934)
Antoine Henri Becquerel (1852–1908)

WHAT COMES NEXT
Harnessing the full potential of nuclear power and nuclear medicine

SEE ALSO
Nuclear fusion and transmutation p.132
Applications of radioactivity p.134
Cold fusion p.176

In 1896, in the wake of the recent discovery of X-rays, Henri Becquerel observed that uranium **atoms** were a source of penetrating radiation. After uranium, thorium was found to be **radioactive.** Pierre and Marie Curie also discovered polonium and radium by observing their radioactive behaviour. The importance of this work is exemplified by the fact that all three won the Nobel Prize for Physics in 1903 for their work. In 1911, Marie Curie was awarded the Chemistry Prize for her discovery of radium and polonium.

Untamed radiation can have a devastating effect on human health, causing sickness and ultimately death. History is littered with the tragic consequences of nuclear accidents and the deliberate use of radioactive materials to harm. The after-effects of the Chernobyl disaster and the murder of Alexander Litvinenko speak to those, but the utilization of radioactivity is also prominent in medicine. Radioactive nuclei can be used to attack diseased cells. Radioactive atoms are also used to trace the movement of atoms in reactions and in radioactive dating, which allows the calculation of the age of ancient artefacts and rocks and minerals.

In 100 words

The spontaneous decay of the nuclei of certain **atoms**, with the associated emission of some combination of high energy electromagnetic radiation and/or charged particles, is known as **radioactivity**. At the end of the 19th century in Paris, France, Pierre and Marie Curie and Henri Becquerel discovered that the radioactivity emitted has varying degrees of penetrating power and has a wide variety of effects on the materials and substances that it comes into contact with. Most commonly, radioactivity consists of three different types: alpha particles (that are positively charged), beta particles (that are negatively charged) and high-energy gamma radiation.

RADIOACTIVITY
Alpha, beta and gamma radiation can be emitted when the nucleus breaks down or is rearranged.

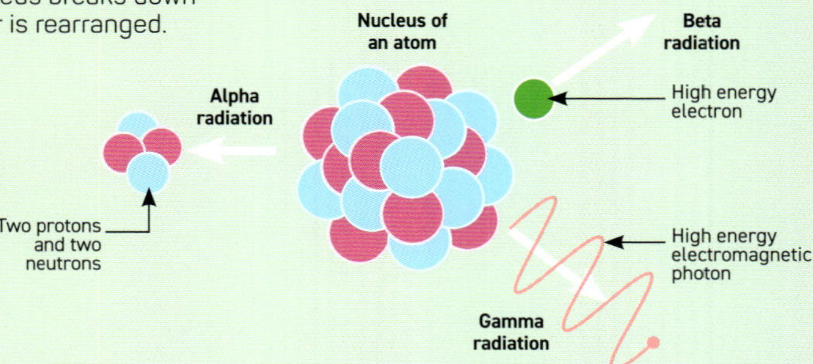

76

Nuclear fusion and transmutation

WHY IT MATTERS
The ability to produce isotopes for research and medical use is a huge boon when naturally occurring elements are not easily available

KEY THINKERS
Ernest Rutherford (1871–1937)
Patrick Blackett (1897–1974)
Ernest Lawrence (1901–1958)

WHAT COMES NEXT
The search for elements beyond 118 is already underway

SEE ALSO
Nuclear fission p.19
Radioactivity p.130
Superheavy elements p.172

Because **transmutation** is both a naturally occurring process and an artificial one, it has a long and complex history. The alchemists' search for the philosopher's stone, to allow for the conversion of base metals into gold (see p.10), is in no way related to modern transmutation at all, but at its heart was a hunt for transmutation. Similarly, the discovery of **radioactivity** was also the discovery of transmutation, although neither of those things would necessarily be thought of in terms of transmutation.

The first artificial transmutation was achieved by Patrick Blackett and Ernest Rutherford in 1919, when nitrogen was converted to oxygen using bombardment with alpha particles. In the early 1930s, Ernest Lawrence invented the first cyclical **particle accelerator** called the cyclotron at the University of California, Berkeley. The invention represented the beginning of the particle accelerator age that allows the artificial production of **isotopes** (see p.20) for research and medical use.

In 100 words

Transmutation is the process of converting one **element** to another via a nuclear reaction. Artificial transmutation involves the bombardment of a target **nucleus** with smaller particles, such as **neutrons**, alpha particles or the nuclei of other elements, in a **particle accelerator**. This allows for the formation of new, superheavy elements (see p.172) when the nuclei fuse together. Natural transmutation occurs during the process of **radioactive** decay and during stellar nucleosynthesis – where element conversion occurs within stars. When the number of **protons** in the nucleus changes, either naturally or artificially, by definition a new element will have been formed.

PARTICLE ACCELERATOR
Particle accelerators bombard a target with high-velocity particles to create new isotopes.

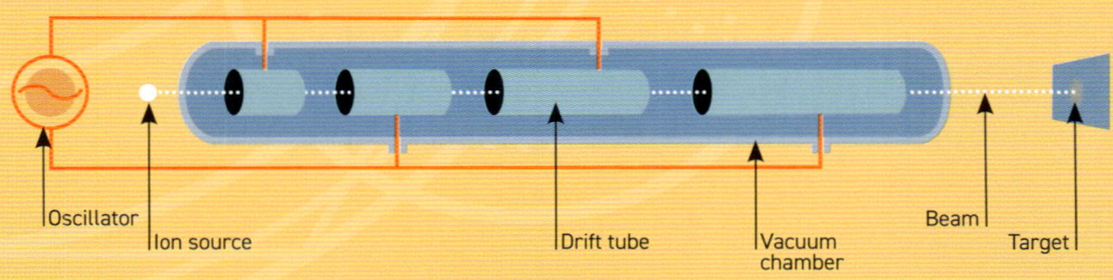

77

Applications of radioactivity

The use of **radioactive** iodine for the treatment of thyroid disorders was pioneered by Saul Hertz and Arthur Roberts during the late 1930s and 1940s. Their work led to the establishment of nuclear medicine. In modern medicine, isotopes such as those of yttrium (Y-90) and lutetium (Lu-177) are used to attack diseased cells. An **isotope** of technetium (Tc-99m) is used in medical imaging in SPECT (Single Photon Emission Computed Tomography).

WHY IT MATTERS
Radioactivity plays a crucial role in modern medicine

KEY THINKERS
Saul Hertz (1905–1950)
Arthur Roberts (1912–2004)

WHAT COMES NEXT
Further research into chemotherapy

SEE ALSO
Radioactivity p.130

In 100 words

Diagnostic testing and the targeting and destruction of diseased cells with radioisotopes remain at the forefront of modern medicine. Various types of radiation are used to sterilize medical equipment and for food preservation. In industry, radiotracers are used to monitor various processes. For example, beta particles can be directed at paper as it is manufactured and the intensity of the radiation passing through the paper can be measured. The results are used to adjust the paper's thickness. Radioisotopes can be used to detect leaks in pipelines, and gamma rays can be used to inspect welds and structural integrity in construction.

Organic Chemistry and Biochemistry

In chemistry, the word "organic" refers to the study of the chemistry of element number six, carbon. Carbon is integral to countless naturally occurring molecules, many of which are essential to life itself, and as such there is strong argument to say that it is the most important of all of the 118 currently known elements. With an ability to catenate (form chemical bonds with itself), it can form a near infinite number of complex chains, rings and cages in combination with other elements.

These organic molecules are what touch us humans the most as they inhabit our bodies, form who we are and control what our bodies do. As the very basis of human life, organic chemistry and biochemistry make us who we are – literally.

78

Organic chemistry and vitalism

WHY IT MATTERS
The realization that organic and inorganic materials are not fundamentally different, and that they can, therefore, all be addressed through the same lens

KEY THINKERS
Jöns Jacob Berzelius (1779–1848)
Friedrich Wöhler (1800–1882)

WHAT COMES NEXT
The explosion of the synthesis of organic compounds

SEE ALSO
Nitrogen-containing organic compounds p.146

By the early 19th century, the success of the synthesis of **compounds** was limited to the inorganic variety, thus reinforcing the idea of vitalism. The famous and brilliant Swedish chemist Jöns Jacob Berzelius was a strong proponent of vitalism, and he argued that organic and inorganic compounds were distinct because of either the presence or absence of the vital force. It wasn't until 1928 that the theory came under significant pressure, but even then it persisted for several years.

In 1928, German chemist Friedrich Wöhler accidentally prepared urea (an organic compound found in urine) from inorganic materials. Wöhler declared to Berzelius that he had made urea "without needing a kidney", meaning without urine! Historians believe that Wöhler's specific role in ending vitalism is somewhat overstated, pointing to many others who poked holes in the theory both before and after his synthesis of urea. They also refute the idea that he was a crusader against vitalism, but there's no doubt that his synthesis of urea is one very important milestone on the way to debunking the theory.

In 100 words

The basic idea behind the philosophical construct of vitalism was that there is a fundamental difference between those things that are considered nonliving objects and those that are living organisms. It assumed that a "vital force" of some kind was present in the latter and absent from the former. When applied to early 19th-century chemistry, this provided a blueprint for the idea that there were two types of chemical substances – those **compounds** that were organic and derived from living things, and those that were inorganic and derived from the earth.

SYNTHESIS OF UREA
The Wöhler synthesis converted ammonium cyanate into urea.

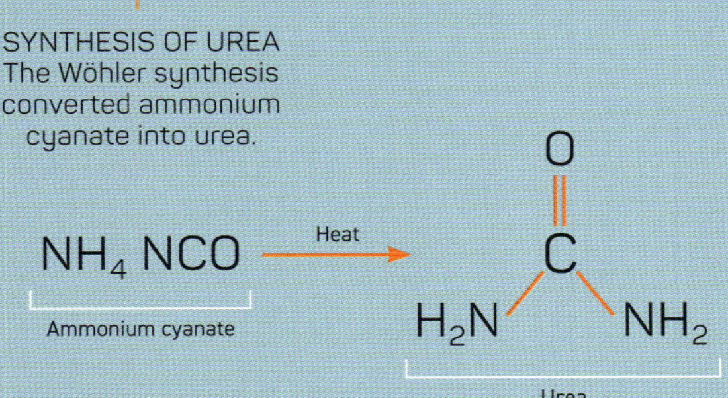

79

Crude oil

WHY IT MATTERS
Innumerable everyday products are sourced from crude oil

KEY THINKER
Eugene Jules Houdry (1892–1962)

WHAT COMES NEXT
The future for fossil fuels remains debatable. The world is still very reliant upon them, so the question becomes: Will alternative energy sources ever fully replace them?

SEE ALSO
Catalysts p.104
Organic chemistry and vitalism p.136
Hydrocarbons p.140
Green chemistry p.174

In fractional distillation, the crude oil mixture is heated, causing vapours to rise up a column. The vapours are repeatedly condensed and re-evaporated, creating a temperature gradient. Depending on the boiling point of the individual components, the **compounds** travel up the column to varying extents. More volatile compounds that have lower boiling points reach the top of the column, and higher boiling point **molecules** condense lower down. Once separated out into "fractions", the various compounds are made into more familiar products, such as petrol, diesel and bitumen.

Crude oil is a complicated mixture of relatively short **hydrocarbon** molecules with carbon chains of two, three and four, mixed in with longer carbon chains. Generally speaking, the shorter chain molecules are more valuable, both for producing important products such as petrol, but also for use in other chemical processes as starting materials. The process of breaking longer hydrocarbon chains into smaller ones is known as cracking. The cracking process can either be thermal or catalytic. Catalytic cracking was pioneered by Eugene Jules Houdry, initially for the purposes of simply producing increased amounts of higher quality gasoline from petroleum. Thermal cracking employs elevated temperatures and pressure to break the carbon chains, while catalytic cracking uses **catalysts**. Catalysts can be expensive, but there can be energy savings since catalytic cracking can run at lower temperatures.

In 100 words

Crude oil is the naturally occurring, black liquid mixture that is usually referred to simply as "oil". It's a fossil fuel, formed over millions of years from organic matter. Oil is primarily a mixture of **hydrocarbons**, with smaller amounts of other **compounds** also present. The specific composition varies widely according to the location and source. Crude oil is separated into its components via fractional distillation – a process that relies upon differences in boiling points of the components. Generally, the smaller chain hydrocarbons are more useful, so to produce them, larger chains are broken up in a process known as cracking.

FRACTIONAL DISTILLATION
The separation of the components of crude oil relies upon the differences in the boiling points of each fraction.

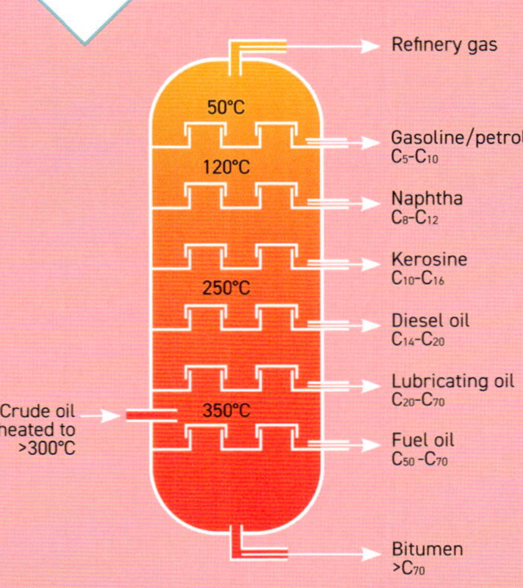

Hydrocarbons

WHY IT MATTERS
The impact of crude oil on the world is hard to overstate, and at its core, crude oil is nothing more than a bunch of hydrocarbons

KEY THINKERS
Friedrich Wöhler (1800–1882)
Friedrich August Kekulé (1829–1896)

WHAT COMES NEXT
The combustion of hydrocarbons and the subsequent production of greenhouse gases, such as carbon dioxide, makes the future uncertain for petrochemicals

SEE ALSO
Combustion reactions p.92
Crude oil p.138
Polymerization p.154

Alkanes that contain carbon **atoms** connected only via single **bonds** are described as saturated, since no further hydrogens can be added to their structures. Alkenes and alkynes on the other hand are unsaturated, since their double and triple bonds would allow the addition of more hydrogen atoms, were the bonds to be broken apart. Arenes, also known as aromatic **hydrocarbons**, contain unsaturated rings of carbon atoms. This characteristic sets them apart from the simpler hydrocarbons, giving them some unique chemistry. The most well-known example of an arene is benzene (C_6H_6).

Carbon's ability to form a multitude of structures, with chains consisting of just one atom to many thousands, together with large rings and three-dimensional structures, means that there are millions of known organic **compounds**. The simpler hydrocarbons have traditionally been used as a source of energy. Their combustion typically releases large amounts of energy, along with carbon dioxide and water.

In 100 words

The chief components of crude oil are **hydrocarbons** – **compounds** formed from only two **elements**, hydrogen and carbon. Carbon's ability to form a total of four **bonds**, both with other **atoms** and with itself – a process known as catenation – is the basis for organic chemistry. There are millions of known organic compounds, with the most common hydrocarbons being families known as alkanes, alkenes, alkynes and arenes. They are characterized by the general chemical formula C_nH_{2n+2}, C_nH_{2n}, C_nH_{2n-2} and C_nH_{2n-6m} respectively, where n is the number of carbon atoms and m is the number of ring structures.

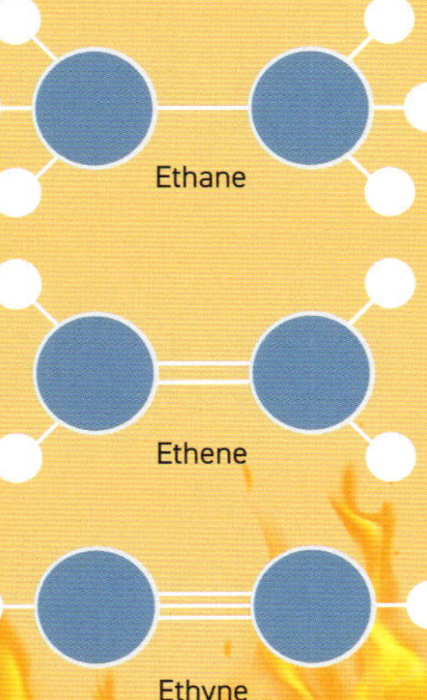

HYDROCARBONS
Ethane, ethene and ethyne are examples of an alkane, an alkene and an alkyne respectively.

81

Alcohols and ethers

Many alcohols and ethers are clear, volatile and highly flammable liquids. Some of them, with smaller **molecules**, are extremely soluble in water (a polar molecule) because of their own polarity, or charged nature. Alcohols have an enormous range of applications in industry and at home, with uses in perfume manufacturing and as disinfectants. Ethers, too, have a wide range of industrial uses in the chemical industry, primarily as solvents for other materials.

WHY IT MATTERS
Around for thousands of years, alcohols and ethers have had an immense influence on society

KEY THINKERS
Taddeo Alderotti (c.1210–1295)
Johann Tobias Lowitz (1757–1804)
Archibald Couper (1831–1892)

WHAT COMES NEXT
The further development of alcohols, such as methanol, and ethers, such as dimethyl ether, as fuels is likely to be important in the future

SEE ALSO
Isomerism p.150
Fermentation p.157

Alcohols and ethers are organic **compounds** that are related since they are structural isomers of one another, with pairs having the same molecular formula. Any given alcohol contains a hydroxyl (-OH) group attached to a carbon **atom**, whereas ethers have an oxygen atom as a bridge between two carbon atoms. Each type of compound contains its own unique functional group – hydroxyl in alcohols and R-O-R in ethers, where R represents a non-carbonyl (not C=O) carbon atom. A common alcohol is ethanol, found in alcoholic drinks, and the most common ether is diethyl ether, a solvent and early anaesthetic.

In 100 words

Aldehydes and ketones are types of organic **compound** that each contain a carbonyl group (-C=O), meaning that the carbon **atom** shares a double **bond** with the oxygen atom. In the case of aldehydes, the carbon atom shares a single bond with at least one hydrogen atom (HC=O), and the carbonyl group appears at the end of a chain, meaning it terminates with a hydrogen atom. In ketones, the carbonyl group appears in the middle of a carbon chain. The carbon atom can share a bond with either carbon or hydrogen atoms (-C=O-). The compounds are often associated with pleasant smells.

82

Aldehydes and ketones

The main uses of aldehydes are in the chemical industry, where they act as precursors to the production of many other compounds. In smaller quantities, but in a more familiar setting, they can also be found as ingredients in some perfume formulations. The vanillin **molecule**, which contributes to vanilla extract aroma, is an aldehyde. Musk deer release a scent rich in muscone, which is a ketone. The Russian-born French perfumer Ernest Beaux used aldehydes in his formulation of the iconic Chanel N°5 in 1920. The most common ketone, acetone, is a household chemical in the form of nail polish remover.

WHY IT MATTERS
These compounds are responsible for many naturally occurring, pleasant aromas

KEY THINKERS
Justus von Liebig (1803–1873)
Leopold Gmelin (1788–1853)

WHAT COMES NEXT
The reactivity of aldehydes and ketones means that there is potential for their use in producing desirable, biodegradable polysaccharides (a class of carbohydrates)

SEE ALSO
Carboxylic acid, esters and derivatives p.144

83

Carboxylic acids, esters and derivatives

WHY IT MATTERS
This collection contains some of the most important and most familiar compounds. For example, vinegar and citric acid are carboxylic acids, and many natural and artificial flavourings contain esters

KEY THINKER
Louis-Jacques Thenard (1777–1857)

WHAT COMES NEXT
Esters are set to be used in coatings for wood and other materials in construction in the future

SEE ALSO
Alcohols and ethers p.142

French chemist Louis-Jacques Thenard worked extensively on ethers (see p.142) and on **compounds** that were originally thought of as ethers but later became recognized as esters. As well as being prominent in nature's aromas and tastes, esters play an important role in artificial perfumery. For example, the famous perfume Chanel N°5 contains benzyl cinnamate – an ester that imparts a sweet, balsamic-like note.

In 100 words

Some of the most common **acids** that you know, such as those in vinegar and in fruit juice, are carboxylic acids. Carboxylic acids contain the functional group -COOH. If you've ever tasted almost any fruit, then you've probably had a bunch of esters in your mouth. Esters can be made from carboxylic acids, often when an acid reacts with an alcohol, and they contain the functional group R-COO-R. Many naturally occurring flavours and smells are esters. Other **molecules** made from carboxylic acids are sometimes known as derivatives of the acids, and they include **compounds** such as acid chlorides and amides.

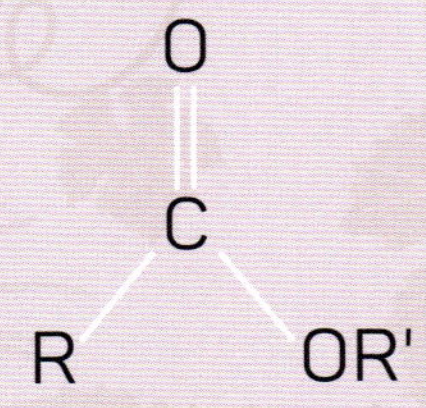

O
C
R OR'

ESTERS
All esters contain a -COO group that connects to R groups of various types. They are responsible for many fruity and fragrant aromas and tastes.

84

Nitrogen-containing organic compounds

WHY IT MATTERS
Nitrogen-containing compounds are used as dyes and extensively for making drugs in the pharmaceutical industry

KEY THINKER
Ascanio Sobrero (1812–1888)

WHAT COMES NEXT
Nitrogen-containing compounds will forever be used in fertilizers, and improvements in them to prevent environmental damage is central to their future, continued use

SEE ALSO
Organic chemistry and vitalism p.136

The various types of organic nitrogen **compounds** are distinguished from one another via their functional groups. A functional group is a distinct group of **atoms** that together have their own, distinct chemical properties. One type of amine includes the -NH2 group. Other types of amines have those hydrogen atoms replaced with carbon chains of various types. Amides are similar to amine but include a carbonyl (-C=O) group, and nitriles have nitrogen atoms triply **bonded** to carbon atoms.

Many nitrogen-containing organic compounds can be used as explosives. TNT (trinitrotoluene) is one example, as is nitroglycerine. The latter was first made by Italian chemist Ascanio Sobrero in 1846. Interestingly it went on to become a powerful medicine, used to open blood vessels for the relief of chest pain. Other well-known organic **molecules** that contain nitrogen atoms are caffeine, cocaine and nicotine. Nitrogen is also a vital **element** in fertilizers (see p.112).

In 100 words

Nitrogen-containing **compounds** include amines, amides, nitriles and azo compounds. Some of those things may not sound all that familiar, but these compounds are prevalent in everyday life. Amines, which contain nitrogen **atoms** connected to at least one carbon atom, are often responsible for terrible smells. The compound trimethylamine is responsible for the smell of rotting fish. The amine **molecules** with the common names cadaverine and putrescine are responsible for the "smell of death", as they are produced by decaying animal flesh. Azo compounds contain a carbon chain with a nitrogen-to-nitrogen double **bond** and are popular dyes.

TNT

TNT, or trinitrotoluene, is a well-known explosive. Compounds containing isolated nitrogen atoms often make good explosives.

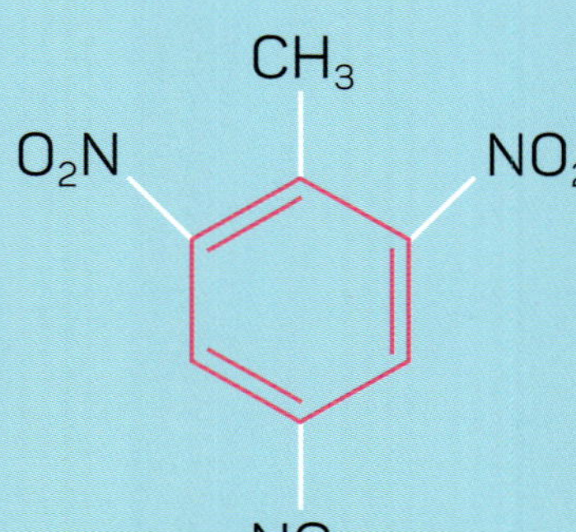

85

CFCs

WHY IT MATTERS
They really don't matter any longer, but CFCs serve to illustrate an important chemical cautionary tale

KEY THINKERS
Frank Sherwood Rowland (1927–2012)
Mario Molina (1943–2020)

WHAT COMES NEXT
Not much! CFCs have been relegated to the dustbin of chemistry due to their environmental impact

SEE ALSO
The halogens p.57

CFCs represent an interesting entry in this book, since they are a group of chemicals that are essentially no longer relevant to chemistry in the modern world. However, there was a time when they were supremely important, and their demise is a fascinating story.

After emerging as refrigerants in the 1940s and 1950s, the 1970s brought about a shocking revelation: in sunlight, CFCs broke down and produced extremely reactive chlorine free radicals. In a chain reaction, the protective ozone layer that prevents harmful UV rays from entering the atmosphere was destroyed by a reaction with these radicals. The problem was found to be so severe that in 1987 the Montreal Protocol essentially outlawed CFCs. The chemists who had brought the problem to the fore, Frank Rowland and Mario Molina, were awarded the Nobel Prize for Chemistry in 1995. Somewhat astonishingly, a single man, Thomas Midgley Jr, was responsible for both the creation of CFCs and the addition of lead in petrol, which were both environmentally disastrous.

In 100 words

CFCs (chlorofluorocarbons), including trichlorofluoromethane (CCl3F) and dichlorodifluoromethane (CCl2F2, also known as freon – a name applied to a number of similar **compounds**) were used as refrigerants starting in the 1940s and 1950s. These compounds were considered safe, and they replaced the toxic refrigerants that had been in use previously. What wasn't known at the time was that CFCs had a propensity to break down in sunlight, producing highly reactive chlorine free radicals. These free radicals have unpaired **electrons** that are ready to react with almost anything, and that's exactly what they did with the ozone layer around Earth.

OZONE PROTECTION
The ozone layer acts as a protective barrier around the Earth, preventing harmful radiation from reaching the atmosphere.

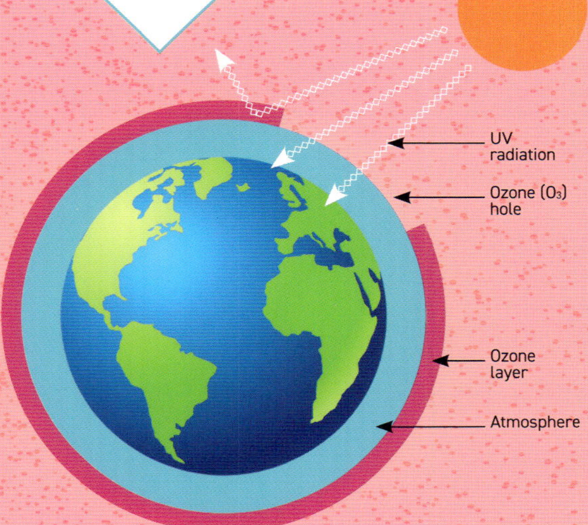

86

Isomerism

WHY IT MATTERS
Given that two isomers of the same compound can have completely different reactions, understanding their distinction from one other is vital

KEY THINKERS
Jöns Jacob Berzelius (1779–1848)
Justus von Liebig (1803–1873)
Friedrich Wöhler (1800–1882)

WHAT COMES NEXT
In 2018, a new type of isomerism involving compounds of boron atoms was discovered, akamptisomerization

SEE ALSO
Covalent bonds p.70

Through his work with French chemist Joseph Louis Gay-Lussac (see p.85), Justus von Liebig succeeded in analysing silver fulminate (AgCNO). While working with Berzelius, Friedrich Wöhler (see p.136) correctly analysed silver cyanate (AgOCN). At the time, the contradictory findings of the two scientists created a serious dilemma – how could two **compounds**, which had several different properties, have the same elemental composition? This contradicted the prevailing theory that chemical properties were directly reliant upon composition. It turns out that isomers of one another – as these two substances are – can have astonishingly different properties even with the simplest rearrangements of **atoms**.

Isomerism is especially crucial in organic chemistry, and in biochemistry, where the interaction of drugs within the body is often reliant upon the three-dimensional positions of those **molecules** in space. The most infamous example is thalidomide – the drug that was touted, and prescribed, as a cure for morning sickness in pregnant women, primarily in the 1950s. Thalidomide exists in two isomeric forms known as optical isomers. One of the isomers worked well in the intended manner, while the other produced terrible birth defects in children. Consequently, the drug was removed from sale in the USA in 1961.

In 100 words

Isomerism is the chemical phenomenon where two **compounds**, which have exactly the same **molecular** formula, have a different physical arrangement of those same **atoms**. Isomerism is commonly divided into two main types: structural isomerism, where the atoms are connected to one another in different sequences, and stereoisomerism, where the connectivity is the same, but the atoms occupy unique and nonequivalent arrangements in three-dimensional space. The Swedish chemist Jöns Jacob Berzelius first formally proposed the idea of isomerism around 1830, but the concept emerged in a somewhat tangled manner following a conflict between two prominent chemists of the time.

(*S*) Thalidomide

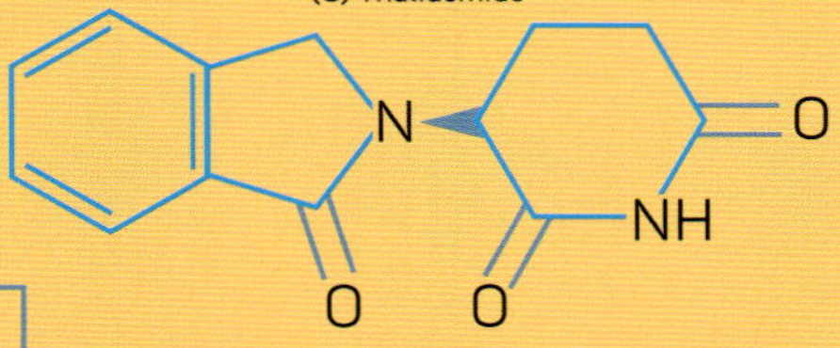

(*R*) Thalidomide

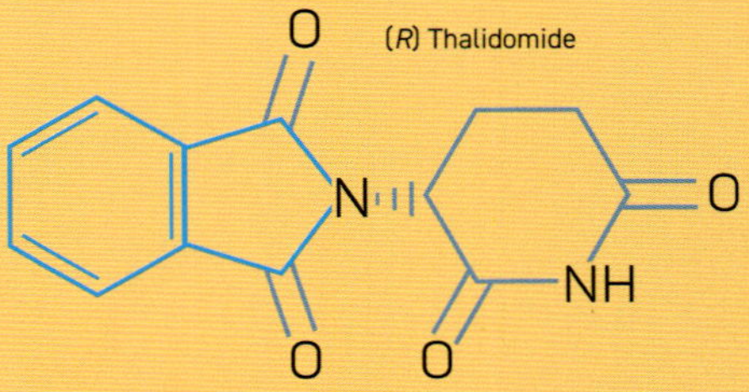

OPTICAL ISOMERS OF THALIDOMIDE
Chiral thalidomide exists in two isomeric forms: the R and the S optical isomers.

87

Chirality

WHY IT MATTERS
Drug therapy is particularly sensitive to chirality, where one molecule does the job and the other can potentially do terrible harm

KEY THINKERS
Frances Oldham Kelsey (1914–2015)
Jacobus Henricus van 't Hoff (1852–1911)

WHAT COMES NEXT
As chirality has such a crucial part to play in biochemistry, you can expect it to be a large part of future drug development

SEE ALSO
Isomerism p.150

The concept of chirality is born out of the three-dimensional arrangement of carbon **atoms** in **molecules**, first championed by van 't Hoff in 1874. Carbon atoms with four distinct groups attached form tetrahedral arrangements and are called chiral centres. Their existence in a molecule leads to what we call optical isomerism. The two forms of the molecule that result are known as enantiomers, and each enantiomer can have its own unique properties. This is especially crucial in biological chemistry and in the chemistry of drugs. Many drugs are chiral **compounds**, with one version of the molecule having the desired therapeutic effect and, in worst case scenarios, the other version being harmful. As mentioned in isomerism (see p.150), perhaps the most devastating example of the problem was demonstrated by the drug thalidomide. Scientist Frances Oldham Kelsey was one of the first medical officers appointed at the US Food & Drug Administration (FDA). Her refusal to grant clearance for thalidomide to be marketed widely in the USA, despite extreme pressure from the manufacturers, saved untold numbers of people from becoming severely disabled.

In 100 words

Some, otherwise apparently identical **molecules**, exist in two forms. They are constructed in such a way that they possess a geometric property which means that their mirror images cannot be superimposed upon one another. This property is called *chirality*. The easiest way to envisage this is to consider your own hands. They are mirror images of one another, but you can never superimpose one on the other. In fact, the two versions of a molecule that possess chirality are sometimes called right and left handed. Chirality occurs in organic molecules when a carbon **atom** has four distinct groups attached to it.

CHIRAL CENTRES
Chiral centres are carbon atoms with four different groups attached to them.

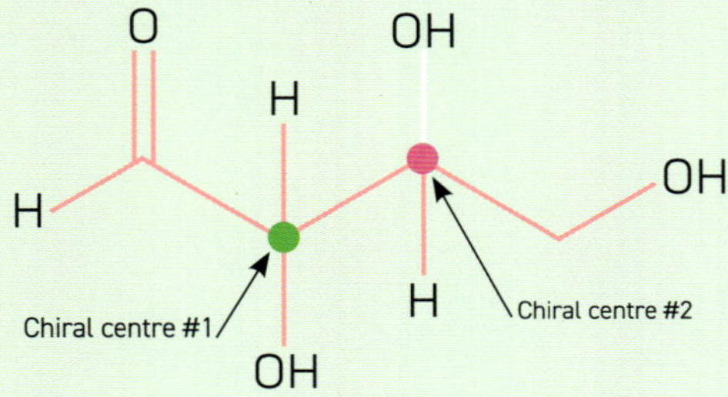

88

Polymerization

WHY IT MATTERS
The revolution in materials science in the 20th century was largely centred around plastics

KEY THINKERS
Hermann Staudinger (1881–1965)
Wallace Carothers (1896–1937)

WHAT COMES NEXT
The development of new polymers is part of the modern search for more advanced materials

SEE ALSO
Crude oil p.138
Hydrocarbons p.140

The importance of polymerization is demonstrated by the many materials and products that have been developed from the process. In particular the whole plastics industry is based upon polymerization. Plastics are synthetic materials that are made primarily from petrochemicals. They are versatile in terms of their properties and therefore their uses revolutionized 20th-century materials science, and frankly, 21st-century life. Their durability, relatively low density, and resistance to moisture and many chemicals, has made them highly sought after but not without cost. They create a significant environmental challenge, mostly due to their inability to biodegrade and associated issues with disposal.

In 100 words

Polymerization is the process that combines huge numbers of smaller **molecules** into a single, long chain molecule. The uncombined smaller molecules are called monomers, and the final long chain molecule is called a polymer. There are two types of polymerization: addition and condensation. In addition polymerization, the monomers join together in a simple chain without producing any by-products. In condensation polymerization, the monomers join together to make a chain, but where the monomers join, they eliminate a small molecule such as water as a by-product. Polymers are ubiquitous in everyday life. Most plastics are polymers, as are familiar synthetic fibres like nylon.

89

Nuclear magnetic resonance

WHY IT MATTERS
NMR has become a completely indispensable tool in chemistry and in medicine in particular

KEY THINKER
Otto Stern (1888–1969)

WHAT COMES NEXT
The portability of smaller NMR machines, and the possibility of their more personalized use in medicine, are set to revolutionize the technique's use in the future

SEE ALSO
IR spectroscopy and UV spectroscopy p.158

NMR finds particular use in determining the structure of organic **molecules**, including large biochemical molecules, such as proteins and nucleic **acid**, and imaging soft body tissue using an MRI machine. One of the chief advantages of NMR is that during the analysis the sample material is not destroyed and can be used again. In medicine, its noninvasive nature is a great boon. Otto Stern won the 1943 Nobel Prize in Physics for his work on determining the magnetic moment of the **proton**.

In 100 words

Nuclear magnetic resonance (NMR) is a powerful spectroscopic technique (see p.158) that examines the spin properties of atomic nuclei. When placed into a strong magnetic field, certain nuclei – most commonly H-1 and C-13 – will align with the applied field. When the nuclei are then exposed to electromagnetic radiation that matches their natural resonance frequency, they absorb the applied energy and they "flip" their orientation in the field. As they "relax" back to their original position, the nuclei release that absorbed energy. The released energy is analysed and can give detailed information about the nature of the **compounds** that contain them.

In 100 words

In the biochemical process known as fermentation, microorganisms such as yeast and bacteria turn carbohydrates (see p.160) into energy. The process takes place in the absence of oxygen and as such is called anaerobic. There are several types of fermentation, one being the creation of alcohol for drinks. In that process, yeast converts sugar into ethanol and carbon dioxide.

$$C_6H_{12}O_6 \rightarrow 2C_2H_5OH + 2CO_2$$

Another type of fermentation that produces lactic **acid** is performed by bacteria and muscle cells converting sugars into lactic acid, which produces that familiar fatigue and "burning" sensation in muscles when vigorous exercise takes place.

Fermentation

Fermentation has been used by humans (either knowingly or unknowingly) for thousands of years. Who knows when the first wine, bread or yoghurt was made? The lactic acid fermentation process is part of the production of foods such as pickles, yoghurt and sauerkraut. Famous French chemist Louis Pasteur first demonstrated experimentally that yeast could turn sugars into alcohol. He believed that fermentation was a type of vitalism – i.e., that only living organisms could be responsible for it.

WHY IT MATTERS
A prehistoric, naturally occurring process, fermentation has been at the heart of human civilization for millennia

KEY THINKER
Louis Pasteur (1822–1895)

WHAT COMES NEXT
Some have proposed the development of what is known as precision fermentation, where microbes can produce specific biochemical molecules in "cell factories"

SEE ALSO
Organic chemistry and vitalism p.136
Carbohydrates and sugars p.160

91

IR spectroscopy and UV spectroscopy

WHY IT MATTERS
Spectroscopy allows analysis of organic compounds without the destruction of the sample

KEY THINKERS
Johann Wilhelm Ritter (1776–1810)
Sir William Herschel (1738–1822)

WHAT COMES NEXT
Smartphones can be used as visible light spectrometers by just about anyone

SEE ALSO
Nuclear magnetic resonance p.156

UV **spectroscopy** (sometimes called UV-vis because the radiation used also spills into the visible part of the spectrum) is particularly important in the analysis of conjugated systems found in larger organic **compounds**. A conjugated system is one where single and double **bonds** are found alternating next to one another in the structure. This is important because so many naturally occurring **molecules** contain such conjugated systems, and UV radiation is particularly sensitive to small changes in these systems, allowing the unique identification of molecules. Most detailed analysis requires the use of both techniques, alongside other methods including mass spectrometry (see p.21) and chemical analysis, where the results of chemical reactions are studied to aid identification. The advantage of spectroscopic analysis is that, unlike in chemical analysis, the samples under test are not destroyed.

In
100 words

Spectroscopy is the study of how matter interacts with electromagnetic radiation of various types. The radiation comes from different parts of the spectrum, including visible light, ultraviolet (UV) and infrared (IR). UV and IR are used chiefly in the analysis of organic **molecules**. IR utilizes the fact that different **covalent bonds** (see p.70) vibrate at specific frequencies when exposed to IR radiation. Analysis produces a unique fingerprint region for **compounds** and distinct, generic absorptions for functional groups allowing for identification. UV spectroscopy uses higher energy radiation than IR, and it causes **electrons** within the molecules to be "excited" by absorbing the radiation.

Carbohydrates and sugars

WHY IT MATTERS
Some of the most important molecules that we know, carbohydrates are the energy source for plants and animals

KEY THINKER
Emil Fischer (1852–1919)

WHAT COMES NEXT
So-called "rare sugars" occur in nature in only very small quantities and have different structures from traditional sugars. They are seen as potential replacements for traditional sugars like glucose and fructose

SEE ALSO
Organic chemistry and vitalism p.136
Hydrocarbons p.140

The chief role of carbohydrates in our world is to act as energy stores and transporters. A great example that you'll be familiar with is glucose ($C_6H_{12}O_6$). Glucose is a monosaccharide, which means that it is a single sugar **molecule**. These small molecules can pair up to produce disaccharides that are also very familiar, for example as sucrose – the sugar we most commonly use in food and cooking ($C_{12}H_{22}O_{11}$). Starch is the energy source for plants, and cellulose is used for their structure, making up a large part of the wood that forms trees.

German chemist Emil Fischer worked on sugars and is best known for his Fischer Projections, which are two-dimensional representations of molecules used extensively with sugars and other carbohydrates. He won the 1902 Nobel Prize in Chemistry for his work with sugars.

SUGARS
One of the most familiar sugars – glucose – and its Fischer projections

In 100 words

It's right there in the name. *Carbohydrate* literally means carbon and water, telling us that the chemical formulae of this vitally important group of chemical substances is centred around C, H and O. Even more specifically, all carbohydrates have the general formula of a C **atom**, coupled with a water **molecule** repeated many times, $C(H_{20})n$. Simple carbohydrates have carbon atoms arranged in ring structures with -OH (hydroxyl) groups attached. The rings can break open to make long chains too. Simple carbohydrates are polar and water-soluble, whereas long chain complex carbohydrates like scratch and cellulose are not.

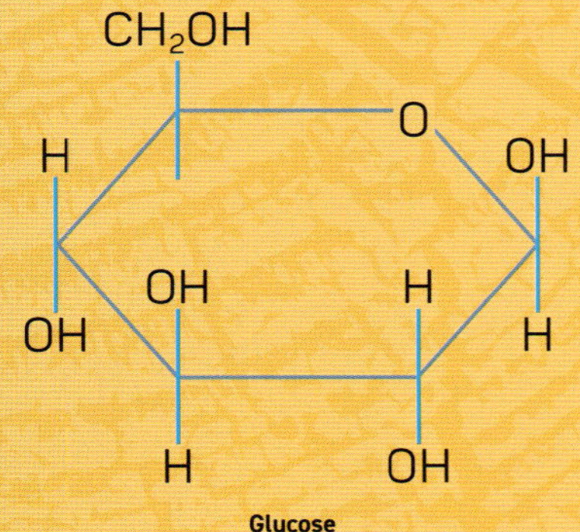

Glucose

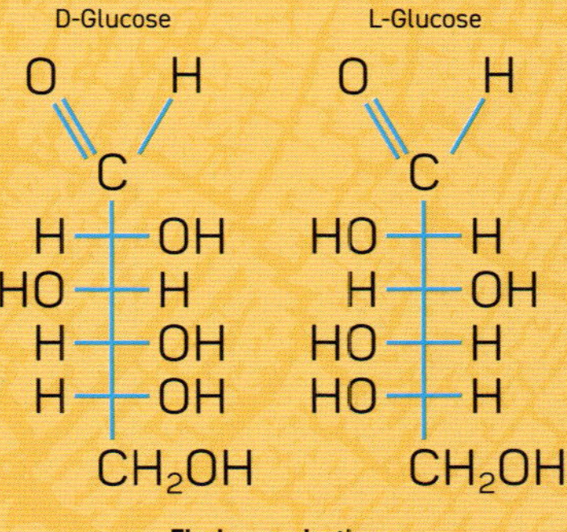

Fischer projections

93

Amino acids and proteins

WHY IT MATTERS
Proteins perform a dizzying array of functions in the human body but are only made from 20 amino acids

KEY THINKERS
Gerardus Johannes Mulder (1802–1880)
Margaret Oakley Dayhoff (1925–1983)

WHAT COMES NEXT
The fascinating and somewhat controversial science of making artificial or cultured meat in laboratories for human consumption is centred around the science of proteins and amino acids

SEE ALSO
Hydrocarbons p.140
Carbohydrates and sugars p.160

Each amino **acid** contains a central carbon **atom** with, appropriately enough, an amino group (-NH_2) and an acid group (-COOH) attached to it. A hydrogen atom is also attached, along with a fourth side chain that is variable in its length and structure. This fourth variable group can impart all kinds of properties to each amino acid, making them acidic, basic, small, large, polar or nonpolar.

Proteins are huge, chain **molecules** that are simply a sequence of amino acids. To describe the sequence, a simple one- or three-letter code is used to represent the amino acids that make them up. Both a three-letter and a one-letter code exist. The one-letter codes were developed by Margaret Oakley Dayhoff, a leading figure in bioinformatics. Gerardus Johannes Mulder was a Dutch chemist who also performed pioneering work on the structure of proteins.

In **100** words

Amino **acids** are the smaller **molecules** that join together in long chains known as proteins. All of the proteins in your body are made up of some combination of 20 of the 500 or so naturally occurring amino acids. Those 20 amino acids can be categorized into two groups: essential and non-essential. Non-essential amino acids can be made by the human body, but essential amino acids have to be consumed via diet. Proteins perform a huge number of functions in the body. They are structural, regulate functions at the cellular level, and they control the speed and efficiency of biochemical processes.

AMINO ACIDS
Amino acids all have the same general structure of -NH_2, -H and -COOH groups, whereas the R group varies from acid to acid.

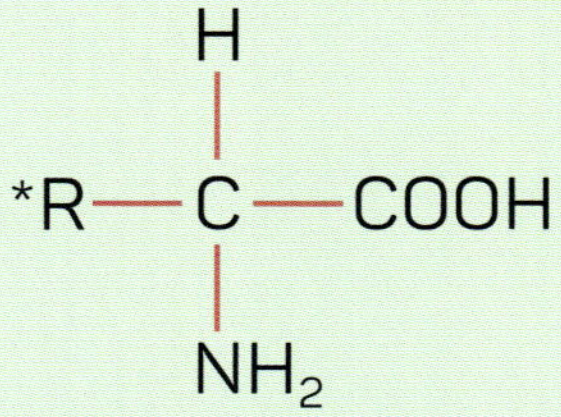

H
|
*R—C—COOH
|
NH_2

94

DNA

WHY IT MATTERS
The very blueprint for life, DNA holds the hereditary information that we pass from generation to generation

KEY THINKERS
Francis Crick (1916–2004)
Maurice Wilkins (1916–2004)
Rosalind Franklin (1920–1958)
James Watson (1928–)

WHAT COMES NEXT
CRISPR (clustered regularly interspaced short palindromic repeats) is a gene-editing tool that allows scientists to modify DNA

SEE ALSO
Intermolecular forces p.76
Polymerization p.154
Amino acids and proteins p.162

DNA molecules have several components: a sugar called deoxyribose, a phosphate and some sequence of organic **bases** that contain nitrogen. The bases are commonly known as A, T , G and C, which represent adenine (A), thymine (T), guanine (G) and cytosine (C) respectively. The double-helix structure resembles a "ladder-like" structure, where sugar and phosphates make up the long sides of the ladder, and bases make up the rungs.

The iconic 1953 discovery of the structure of DNA by Watson and Crick, and the subsequent awarding of the 1962 Nobel Prize in Physiology or Medicine for the same, has become almost as famous for its failure to properly recognize the role of British female chemist Rosalind Franklin. Her data was used in the discovery without her knowledge. She died in 1958. Franklin became a feminist cause célèbre as a result of her snub, although interestingly, Maurice Wilkins, who was also mostly left out initially, was never lamented to the same degree, probably in part because he was ultimately recognized via the Nobel.

In 100 words

The abbreviation "**DNA**" has entered the vocabulary of just about everybody who has ever taken an interest in any type of police or crime-based show on television. Deoxyribonucleic **acid**, in its classic, double-helix arrangement, is the **molecule** that determines human traits, such as the colour of our eyes and hair, and it passes those traits to offspring. Your unique DNA is inherited from your parents, which is why you carry traits that you can recognize in your mother, father and other relatives. The relatively simple and rather beautiful structure carries the genetic information needed for life.

DNA
The classic double-helix arrangement of DNA contains the genetic information inherited from your parents.

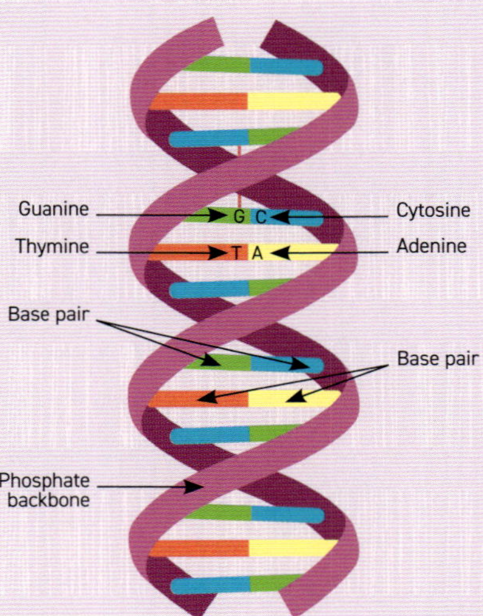

Modern Chemistry and the Future

The very nature of scientific discovery means that it is never ending. The constant quest for more knowledge, greater understanding and solutions to contemporary problems drives chemistry forward on a daily basis.

As boundaries are pushed and progress is made, mistakes and setbacks are inevitable. New scientific discoveries can come with a plethora of failure as scientists continue to probe the edges of what is currently possible. Will cold fusion ever be possible? Can we save the planet from destruction? How will computer technology and AI shape chemistry going forward? How far can we expand the periodic table? All of these questions can only be answered by the chemistry of the future.

In 100 words

Computational chemistry is a field of chemistry that utilizes computer technology to make predictions about reactions and **molecules** before they have been performed or produced. Software simulates chemical reactions and makes as yet unproven structures visual. Its chief role is to allow the prediction of multiple outcomes before any chemicals are mixed together, often by analysing large data sets. It helps greatly in the design of new drugs and in understanding complicated reaction mechanisms. Mathematical **models** provide a window into processes and molecules that would otherwise be difficult or impossible to reach via traditional experimentation.

95

Computational chemistry

The field of computational chemistry is a relatively new one. As computing power has become increasingly utilized and understood, more and more applications have been found in the physical sciences. A watershed moment in the history of computational chemistry was the 1998 Nobel Prize recognition of the theoretical chemists Walter Kohn and John Pople. They received the prize for their work with computational methods in quantum theory.

WHY IT MATTERS
A virtual laboratory is an invaluable tool that enhances safety, saves time and money and speeds up the research process

KEY THINKERS
Walter Kohn (1923–2016)
John Pople (1925–2004)

WHAT COMES NEXT
As Artificial Intelligence continues to advance, there appear to be limitless possibilities for the prediction of new processes, molecules and the analysis of data

SEE ALSO
Green chemistry p.174

96

Quarks and the Standard Model

WHY IT MATTERS
Quarks are one part of the Standard Model of particle physics, which in turn explains how the universe is made of just a few fundamental particles and forces

KEY THINKERS
Murray Gell-Mann (1929–2019)
George Zweig (1937–)

WHAT COMES NEXT
The Standard Model still has some limitations, and further research is always driving it forward, as evidenced by the discovery of the Higgs boson in 2012

SEE ALSO
Chadwick and the neutron p.17
The nuclear model p.24
The quantum mechanical model p.28

Starting in the 1960s, the **quark model** was developed independently by Murray Gell-Mann and George Zweig. **Protons** and **neutrons** are collectively known as hadrons, and are composite particles, each being made up of the elementary quarks. These hadrons are held together by what physicists call the "strong **nuclear force**". Since energy and mass are interchangeable via Einstein's famous equation $E = mc^2$, it is this force, via its energy, that contributes around 99 per cent of the mass of any atomic **nucleus**. The mass of quarks themselves is negligible in comparison to the mass that comes from the nuclear force. As for **electrons**, in the **Standard Model** of nuclear physics, electrons are still thought of as fundamental particles and as such are not made of anything smaller.

Up and down quarks are just two types of fundamental particles in the Standard Model. Not only are there other types of quark, there are also other particles, collectively known as **leptons**, and multiple force carriers.

STANDARD MODEL OF ELEMENTARY PARTICLES
This diagram summarizes the different types of elementary particle that make up the Standard Model.

In 100 words

There was a time when **protons**, **neutrons** and **electrons** were thought to be the smallest particles that make up **atoms**. Now we know that **quarks** are **elementary particles** that make up protons and neutrons. There are six types of quark, and two of them (up and down) make up a proton. The proton is made from 2 up quarks, with a charge of +⅔ each, and 1 down quark, with a charge of -⅓. Together they give the proton a charge of +1. The neutron is made of 1 up quark and 2 down quarks, meaning its total charge is 0.

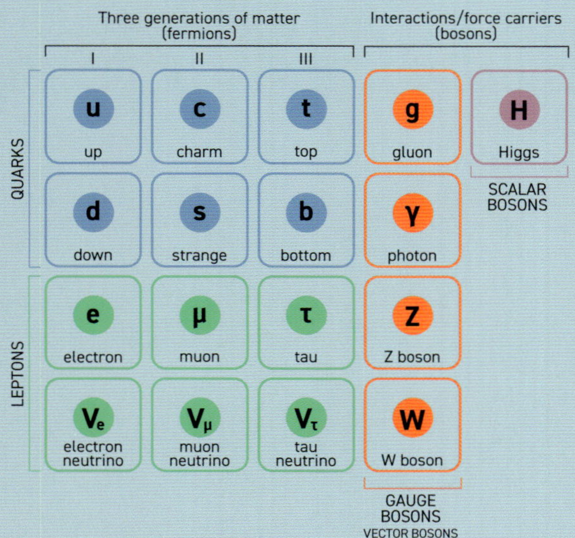

97

Nano-technology

WHY IT MATTERS
Nanotechnology pushes the boundaries of what we thought was possible by observing the macro world alone, as properties of materials at the atomic and molecular level are so different

KEY THINKERS
Gerd Binnig (1947–)
Heinrich Rohrer (1933–2013)

WHAT COMES NEXT
Nanotechnology has seemingly limitless possibilities in multiple fields, including the use of nanorobots in single cell manipulation

SEE ALSO
The quantum mechanical model p.28

Hydrogen **atoms** are the smallest atoms. They have diameters of approximately 0.1 nanometers. The largest atoms have diameters of around 0.5 nanometers. That means that the combination of just a few atoms in a three-dimensional structure or **molecule** fulfils the criteria for **nanotechnology**. Nanotechnology is literally performed at the atomic and molecular level.

The scanning tunnelling microscope (STM) represents a huge step forward in nanotechnology. Invented in 1981 by Gerd Binnig and Heinrich Rohrer, it uses a sharp metal tip close to a conductive surface and measures the electrical current that flows between them. This has the effect of mapping the surface of a material atom by atom. The importance of the invention was acknowledged with the award of the Nobel Prize in Physics in 1986.

The incredibly small size of atoms complicates the properties that are observed and predicted at the larger scale via what are known as relativistic effects. These effects are those that consider general relativity in a way that quantum mechanics doesn't, leading to unexpected properties. On such a small scale, there is an unusually high surface-to-volume ratio of the material, meaning that the surface atoms have an exaggerated role in determining properties – this is known as surface dominance. If nanomaterials are known to act in significantly different ways to similar bulk materials, this opens up a whole world of possible applications.

In
100
words

Nanotechnology is the study and manipulation of matter with at least one dimension that possesses a size of between 1 and 100 nanometers. The *nano* prefix is used in the International System of Units and comes from the Greek *nânos*, meaning "dwarf". *Nano* has a precise meaning, i.e., that a factor of 10_{-9} is applied to the unit. For example, a measurement of between 1 and 100 nanometers means a distance of between 0.000000001 m and 0.000000100 m. At such a small scale – the nanoscale – the properties of matter can become very different to that of the visible world.

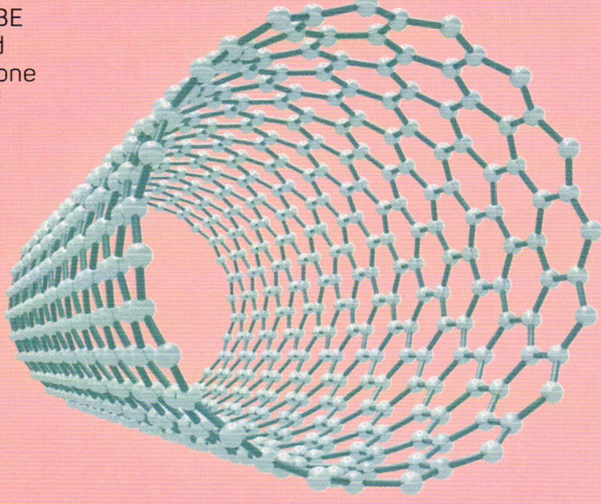

CARBON NANOTUBE
This single-walled carbon nanotube is one allotropic form of carbon.

Superheavy elements

WHY IT MATTERS
Pushing the boundaries of science is the very essence of science itself

KEY THINKERS
Albert Ghiorso (1915–2010)
Yuri Oganessian (1933–)
Peter Armbruster (1931–2024)

WHAT COMES NEXT
How far can the periodic table be extended? That's the question for the superheavy chemists

SEE ALSO
The actinides p.62
Nuclear fusion and transmutation p.132

The question of how far the **periodic table** can be extended is fraught with both theoretical and practical barriers. Various scientists have weighed in on the question, notably Richard Feynman, who predicted a limit at **atomic number** 137. More recent predictions go as far as **element** number 173. Not only is the stability of such **atoms**, and, therefore, their very existence debatable, but there are also limitations on the technology used to create and detect the nuclides themselves. In relation to stability, the theoretical island of stability refers to a collection of superheavy nuclides that have unusually long half-lives compared to similar elements. These elements are said to have "magic numbers" of **protons** and **neutrons** that aid their stability. Magic numbers remain a difficult thing to predict with any certainty, but there is some evidence for 184 being a magic neutron number. Element 114 (flerovium) has been studied as a potential candidate for heavier, more stable **isotopes** because it has demonstrated some longer half-lives in some instances.

The USA, Russia and Germany have been at the forefront of research into and the synthesis of new superheavy elements, and scientists including Albert Ghiorso, Yuri Oganessian and Peter Armbruster have made significant contributions. Oganessian's work was recognized with the naming of element 118 (oganesson), and the others have also contributed to multiple discoveries of new elements.

In 100 words

The superheavy **elements** are generally thought of as those elements with **atomic numbers** greater than 103, but no formal definition exists. Using that guideline, they are currently composed of elements 104 (rutherfordium) up to 118 (oganesson). All of them are artificially produced in laboratories by crashing smaller nuclei into one another in fusion reactions (see p.132). They exist only in tiny quantities – just a few **atoms** at a time – and they are all short-lived, **radioactive** elements, often with half-lives of only fractions of seconds. These elements are of interest purely in theoretical terms, with no known practical applications as yet.

99

Green chemistry

WHY IT MATTERS
Environmental wellness and sustainability depend upon the continued development of green chemistry

KEY THINKERS
Rachel Carson (1907–1964)
Paul Anastas (1962–)
John Warner (1962–)

WHAT COMES NEXT
As demand increases for Earth's natural resources, particularly in areas such as battery technology, green chemistry will become even more important

SEE ALSO
Computational chemistry p.167

Rachel Carson was a marine biologist who single-handedly, but perhaps unwittingly, launched the environmental movement in the USA with a simple action. In 1962 she wrote the book *Silent Spring* that documented the impact of several chemicals, notably DDT, on the environment, and environmental awareness on a large scale was born.

In 1998 Paul Anastas and John Warner created a framework of 12 guiding principles of **green chemistry**, which included minimizing waste rather than working on methods to clean it up, and efficiency via **atom** economy, where the most desirable methods make sure that as much of the starting materials as possible ends up in the desired, final product.

In teaching, the relatively recent advent of microscale chemistry has contributed to the green movement, as well as increasing safety in the educational setting. Rather than using sizable volumes of chemical **solutions** in large beakers during reactions, the emphasis has shifted to small well-plates, with (literally) drops of chemicals being used to observe the same chemical reactions. This not only reduces the use of resources but also makes clean-up easier and promotes a more sustainable laboratory environment.

In 100 words

Green chemistry aims to reduce the environmental impact of the chemical industry by adopting processes and producing chemicals that minimize the use of toxic materials and produce fewer waste products. Sustainability is a major consideration of green chemistry, which seeks more efficient production methods and the use of fewer raw materials. The use of water as a solvent rather than organic substances is central, as are energy-efficient processes. Taken together, these ideals are designed to protect Earth but also continue to allow the growth of the chemical economy. As such, green chemistry is not a political movement.

"Only within the moment of time represented by the present century has one species – man – acquired significant power to alter the nature of the world."

Rachel Carson, *Silent Spring*

100

Cold fusion

WHY IT MATTERS
If there could ever be such a thing, some argue that the energy needs of the world would be solved for good

KEY THINKERS
Stanley Pons (1943–)
Martin Fleischmann (1927–2012)

WHAT COMES NEXT
The idea of cold fusion, while certainly isolated at the periphery, remains a tantalizing possibility for a small group of people who continue to look into the possibility of it being a reality

SEE ALSO
Nuclear fission p.19
Nuclear fusion and transmutation p.132
Superheavy elements p.172

The idea of cold fusion first emerged in 1989, when Stanley Pons and Martin Fleischmann claimed to have achieved such a reaction using palladium electrodes in heavy water at the University of Utah. However, attempts to replicate the research failed, and swiftly the whole concept was dismissed. So why does the idea of cold fusion still linger? Because if it were ever to become a reality, it could provide a limitless supply of clean and renewable energy. As recently as 2019, an article in the scientific journal *Nature* reported that a Google-backed programme had been investigating the theory, but it, too, had failed to achieve the desired results.

In 100 words

The hypothetical idea of cold fusion describes a nuclear fusion reaction that occurs at low (room) temperatures. Under normal circumstances, the fusion of two positive nuclei requires a huge amount of energy to overcome the repulsive force that exists between two similarly charged entities. Cold fusion is a unique topic in this book: all of the other 99 ideas are accepted as valid scientific facts and theories, but cold fusion has no accepted theoretical basis in science. Still, small groups of researchers, mostly at the fringes of the mainstream, continue to investigate and probe its possibilities.

MODERN PERIODIC TABLE OF ELEMENTS

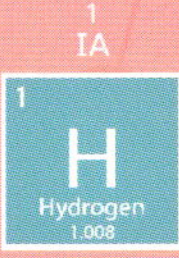      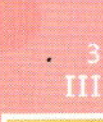           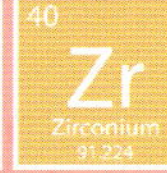           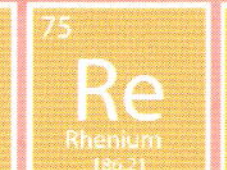    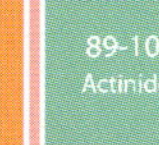     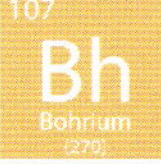   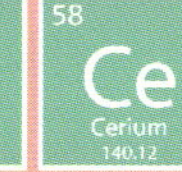   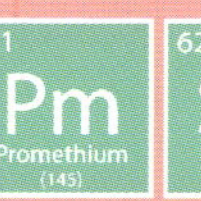 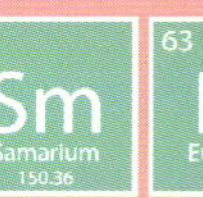      

1 IA	2 IIA	3 IIIB	4 IVB	5 VB	6 VIB	7 VIIB	8 VIIIB	
1 H Hydrogen 1.008								
3 Li Lithium 6.94	4 Be Beryllium 9.0122							
11 Na Sodium 22.990	12 Mg Magnesium 24.305							
19 K Potassium 39.098	20 Ca Calcium 40.078	21 Sc Scandium 44.956	22 Ti Titanium 47.867	23 V Vanadium 50.942	24 Cr Chromium 51.996	25 Mn Manganese 54.938	26 Fe Iron 55.845	27
37 Rb Rubidium 85.468	38 Sr Strontium 87.62	39 Y Yttrium 88.906	40 Zr Zirconium 91.224	41 Nb Niobium 92.906	42 Mo Molybdenum 95.95	43 Tc Technetium (98)	44 Ru Ruthenium 101.07	45
55 Cs Caesium 132.91	56 Ba Barium 137.33	57-71 Lanthanides	72 Hf Hafnium 178.49	73 Ta Tantalum 180.95	74 W Tungsten 183.84	75 Re Rhenium 186.21	76 Os Osmium 190.23	77
87 Fr Francium (223)	88 Ra Radium (226)	89-103 Actinides	104 Rf Rutherfordium (267)	105 Db Dubnium (268)	106 Sg Seaborgium (269)	107 Bh Bohrium (270)	108 Hs Hassium (277)	109

57 La Lanthanum 138.91	58 Ce Cerium 140.12	59 Pr Praseodymium 140.91	60 Nd Neodymium 144.24	61 Pm Promethium (145)	62 Sm Samarium 150.36	63
89 Ac Actinium (227)	90 Th Thorium 232.04	91 Pa Protactinium 231.04	92 U Uranium 238.03	93 Np Neptunium (237)	94 Pu Plutonium (244)	95

18
VIIIA
13
IIIA
14
IVA
15
VA
16
VIA
17
VIIA
11
IB
12
IIB
2 He Helium 4.0026
5 B Boron 10.81
6 C Carbon 12.011
7 N Nitrogen 14.007
8 O Oxygen 15.999
9 F Fluorine 18.998
10 Ne Neon 20.180
13 Al Aluminium 26.982
14 Si Silicon 28.085
15 P Phosphorus 30.974
16 S Sulphur 32.06
17 Cl Chlorine 35.45
18 Ar Argon 39.948
29 Cu Copper 63.546
30 Zn Zinc 65.38
31 Ga Gallium 69.723
32 Ge Germanium 72.630
33 As Arsenic 74.922
34 Se Selenium 78.971
35 Br Bromine 79.904
36 Kr Krypton 83.798
47 Ag Silver 107.87
48 Cd Cadmium 112.41
49 In Indium 114.82
50 Sn Tin 118.71
51 Sb Antimony 121.76
52 Te Tellurium 127.60
53 I Iodine 126.90
54 Xe Xenon 131.29
79 Au Gold 196.97
80 Hg Mercury 200.59
81 Tl Thallium 204.38
82 Pb Lead 207.2
83 Bi Bismuth 208.98
84 Po Polonium (209)
85 At Astatine (210)
86 Rn Radon (222)
111 Rg Roentgenium (282)
112 Cn Copernicium (285)
113 Nh Nihonium (286)
114 Fl Flerovium (289)
115 Mc Moscovium (290)
116 Lv Livermorium (293)
117 Ts Tennessine (294)
118 Og Oganesson (294)
65 Tb Terbium 158.93
66 Dy Dysprosium 162.50
67 Ho Holmium 164.93
68 Er Erbium 167.26
69 Tm Thulium 168.93
70 Yb Ytterbium 173.05
71 Lu Lutetium 174.97
97 Bk Berkelium (247)
98 Cf Californium (251)
99 Es Einsteinium (252)
100 Fm Fermium (257)
101 Md Mendelevium (258)
102 No Nobelium (259)
103 Lr Lawrencium (266)

Glossary

Absolute zero: the lowest temperature possible when all atomic movement has its minimum value. Absolute zero is 0 Kelvin, −273.15 Celsius and −459.67 Fahrenheit.

Acid: a substance that releases hydrogen ions ($H+$) and has a pH level of less than 7.

Acidic strength: determined by its ability to donate hydrogen ions, acids whose molecules donate 100 per cent of their hydrogens are said to be strong, while those that donate much smaller percentages are said to be weak.

Acidic concentration: a measure of how diluted an acid is – large amounts of acid dissolved in small amounts of water are said to have high concentrations, and vice versa.

Activation energy: the minimum amount of energy required to allow a certain reaction to take place.

Alchemy: a protoscience primarily focused on creating gold and the elixir of life.

Allotrope: the atoms of an element bonded in different ways – the way that atoms are arranged and bonded changes its properties.

Atom: the most basic particles of elements, made up of protons, electrons and neutrons.

Atomic mass/weight: a relative figure of the mass of an atom, made up by the combined mass of the protons and neutrons in the nucleus.

Atomic number: the number of protons in an atom's nucleus (also called the proton number). Each element has a different atomic number.

Base: the chemical opposite of an acid, with a pH level greater than 7, that accept hydrogen ions, produce hydroxide ions ($OH-$) in solution and can act as electron pair donors.

Bonds: an attractive force between atoms or molecules that holds them together, made or broken by chemical reactions.

Catalyst: a substance that speeds up a chemical reaction by providing an alternative pathway for the reaction that has a lower activation energy.

Chemical equation: shows the reactants and the products represented by chemical formulae with an arrow to show the direction of the reaction. They must show the same number and the same types of atoms on each side of the equation.

Chemical periodicity: the observation that the properties of the elements and their compounds can often be accurately predicted by the relative positions of the elements in the periodic table.

Compound: a molecule made of different types of atoms.

Covalent bond: a chemical bond in which atoms share electrons, increasing their stability by completing their octects of electrons.

Crystal: a solid structure in which molecules form an organised and repeating pattern.

Diatomic: molecules made of only two atoms.

DNA: large molecule made of (usually) two strands of nucleotides with bases A, C, G and T.

Electron: a negatively charged subatomic particle that, along with protrons and neutrons, make up atoms.

Electronegativity: the measure of how easily an atom can attract electrons to itself within a covalent bond, measured on the Pauling scale.

Elementary particle: a particle such as the electron which is thought to have no underlying structure.

Element: a chemical substance whose atoms all have the same atomic number and that cannot be broken down by a chemical reaction to produce anything simpler.

Endothermic reaction: a reaction that absorbs energy from its surroundings, making it feel cooler.

Entropy: a measure of the randomness in an arrangement of particles.

Exothermic reaction: a reaction that releases energy into its surroundings, making it feel warmer.

Enzyme: biological catalysts that make chemical reactions in the cell go faster.

Four elements: the ancient Greek idea that all matter is made from air, fire, water and earth.

Free electrons: electrons in a metal which are not attached to particular atoms so that they can move within the bulk of the metal.

Fullerene: an allotrope of carbon bound together by covalent bonds.

Green chemistry: a form of chemistry that aims to reduce the environmental impact of the chemical industry by adopting processes and producing chemicals that minimize the use of toxic materials and produce fewer waste products.

Heteronuclear: a molecule consisting of multiple types of atom– i.e., glucose, ($C_6H_{12}O_6$).

Homonuclear: a molecule consisting of only one type of atom– i.e., oxygen (O_2).

Hydrocarbon: an organic compound consisting entirely of hydrogen and carbon, which releases large amounts of energy when burned and therefore is often used as fuel. However, burning them produces carbon dioxide and therefore is extremely dangerous to the environment.

Ionic bond: the electrostatic attraction between positive and negative ions – one atom loses negative electrons to become positive, and passes that negative electron to another atom that then itself becomes negative.

Ions: atoms or groups of atoms which are charged because they have gained or lost electrons.

Isotopes: atoms of the same element (the same number of protons) but with differing numbers of neutrons, giving them varying atomic masses.

Leptons: elementary particles such as electrons and neutrinos and which, unlike quarks, do not experience the strong nuclear force.

Metallic bond: the bond that holds metal elements together through the electrostatic attraction between positive metal ions in a sea of negative electrons.

Model: a simplified version of a physical situation, devised to explain observations and make predictions.

Mole: the amount of substance containing the same number of particles as there are atoms in 12 grams of carbon-12. This number, known as Avogadro's number or Avogadro's constant, is equal to 6.022×10^{23}.

Molecule: an electrically neutral entity consisting of more than one atom.

Nanotechnology: the study and manipulation of matter with at least one dimension that possesses a size of between 1 and 100 nanometers. At such a small scale the properties of matter can become very different to that of the visible world.

Neutron: a neutrally charged subatomic particle discovered by James Chadwick, that along with protons, make up the nucleus of every atom but hydrogen.

Noble gases: the unreactive elements found in Group 18, also known as the inert gases.

Nuclear fission: The formation of lighter atomic nuclei when a heavy nucleus splits apart.

Nuclear force: One of two fundamental forces, weak or strong, which act between particles in the atomic nucleus.

Nuclear model: a model of the atom informed by the experiments of Ernest Rutherford, showing a dense, positively charged centre of mass.

Nucleus: the positively charged central core of an atom, having most of the atom's mass.

Oxidation reaction: a reaction in which electrons are lost.

Particle accelerators: equipment that bombards a target with high-velocity particles to create new isotopes.

Periodic table: the iconic table that arranges elements by increasing atomic number and places chemically similar elements in vertical columns, known as groups, and elements with gradually changing characteristics in horizontal rows, called periods. The relative position of each element allows predictions to be made about its behaviour and properties.

Photosynthesis: a process used by plants and algae to convert sunlight and carbon dioxide into sugars and oxygen and an example of an organic redox reaction.

pH scale: a scale, usually ranging from 0 to 14, that classifies a solution as being acidic, basic or neutral.

Plum pudding model: a model of the atom proposed by J J Thomson showing negatively charged electrons dispersed in a cloud of positive charge.

Polyatomic: a molecule consisting of multiple atoms.

Proton: a positively charged subatomic particle that, along with electrons and neutrons, makes up atoms.

Quarks: elementary particles which experience the strong nuclear force so that they combine to form particles such as protons and neutrons.

Radioactivity: the spontaneous decay of the nuclei of certain atoms, with the associated emission of some combination of high energy electromagnetic radiation and/or charged particles.

Reduction reaction: a reaction in which electrons are gained.

Redox reaction: a reaction in which both oxidation and reduction take place – these two processes are reciprocal to one another as the electrons gained in the reduction process are the same electrons lost in the oxidation process.

Solution: when one substance (the solute) completely dissolves in another substance (the solvent) producing a homogeneous mixture.

Stoichiometric coefficients: the numbers that appear in front of each chemical formulae, which indicate the ratio of the moles of each species in a reaction and must be balanced.

Strand: one chain of a DNA or RNA molecule. Double strands are connected by matching base pairs A-T (or A-U) and C-G.

SI system (Système Internationale): the system of units used in science, based on fundamental units such as the metre, kilogram and second.

Spectroscopy: the study of how matter interacts with electromagnetic radiation of various types.

Standard model: the set of elementary particles which combine to form all of matter, together with the force between them.

Transmutation: the process of converting one element to another via a nuclear reaction. When the number of protons in the nucleus changes, either naturally or artificially, by definition a new element will have been formed.

Vacuum: a space entirely empty or devoid of matter.

Index

D

E

F

G

H

About the author

Adrian Dingle is a communicator, educator and author with more than three decades of experience in conveying and explaining chemistry to various audiences. He is the creator of the award-winning chemistry website *Adrian Dingle's Chemistry Pages* at **www.adriandingleschemistrypages.com**.

He holds a BSc. (Hons.) Chemistry, and a Postgraduate Certificate in Education (SecondaryChemistry), both from the University of Exeter in England.

In addition to writing *Chemistry: 100 Ideas in 100 Words*, he has written several other children's and popular chemistry titles that include *The Periodic Table: Elements With Style*, *How To Make A Universe With 92 Ingredients*, and *DK's Eyewitness Periodic Table*. He is the 2011 winner of the School Library Association of the UK's Information Book Award and in 2012 was honored with the prestigious literary prize *Wissenschaftsbuch des Jahre*, sponsored by the Austrian Ministry of Science and Research. He also adapted Sam Kean's *New York Times* bestseller *The Disappearing Spoon* for young readers. In 2015 he was elected as the inaugural High School Ambassador on the Governing Board of the American Association of Chemistry Teachers (AACT).

He currently teaches at Canterbury School in Indiana. He previously taught at The Culver Academies (also in Indiana) and in Atlanta, Georgia at The Westminster Schools. He lives in Indiana, USA where he continues to follow his beloved Leeds United from afar.

In association with the Science Museum

The Science Museum is part of the Science Museum Group, the world's leading group of science museums that share a world-class collection providing an enduring record of scientific, technological and medical achievements from across the globe. Over the last century the Science Museum, the home of human ingenuity, has grown in scale and scope, inspiring visitors with exhibitions covering topics as diverse as robots, code-breaking, cosmonauts and superbugs. **www.sciencemuseum.org.uk.**

Picture Credits

The publisher would like to thank the following for their kind permission to reproduce their photographs:
(Key: a-above; b-below/bottom; c-centre; f-far; l-left; r-right; t-top)

1 123RF.com: Bill2499 (Background). **2-3 Dreamstime.com:** Alena Ohneva (Background). **4-5 Shutterstock.com:** Beliavskii Igor (Background). **8 123RF.com:** Bill2499 (Background). **10 Adobe Stock:** Varvara_Iur (bl). **19 Alamy Stock Photo:** Horizon International Images (Background). **25 Shutterstock.com:** Rktz (b). **30 Shutterstock.com:** Nadia Murash. **37 Shutterstock.com:** Abdullaharsln (b). **38-39 Dreamstime.com:** Oksana Pasishnychenko. **39 Shutterstock.com:** Abdullaharsln (b). **47 Shutterstock.com:** CESM I Studio (b). **50 Adobe Stock:** JL-art (b). **51 Shutterstock.com:** CESM I Studio (b). **55 Adobe Stock:** Tollaru (br). **56 Dreamstime.com:** Alena Ohneva (Background). **59 Shutterstock.com:** CESM I Studio (b). **61 Shutterstock.com:** CESM I Studio (b). **63 Shutterstock.com:** CESM I Studio (b). **65 Adobe Stock:** Pro500 (Background). **70-71 Adobe Stock:** Sergey Shimanovich (Background). **76-77 Shutterstock.com:** Patricia F. Carvalho (Background). **85 Shutterstock.com:** Yaruna (Background). **90-91 Adobe Stock:** Quirk Craft Studio (Background). **94 Dreamstime.com:** Stanislav Bokser (Background). **100-101 123RF.com:** Studiom1 (Background). **116-117 Adobe Stock:** Good Job (t). **118-119 Adobe Stock:** LeonART (b). **127 Adobe Stock:** Nandalal (b). **128-129 Dreamstime.com:** Alena Ohneva (Background). **132-133 Shutterstock.com:** Vonzur (Background). **135 Dreamstime.com:** Viktoriia Kotova (Background). **140-141 Dreamstime.com:** Sensvector (b). **144-145 Shutterstock.com:** Pongsapol Ponata (Background). **148-149 Shutterstock.com:** Kindlena (Background). **149 Adobe Stock:** MilletStudio (b). **154-155 Dreamstime.com:** Svsunny (Background). **160-161 Shutterstock.com:** Natasha Barsova (Background). **164-165 Dreamstime.com:** Aenota (Background). **165 Shutterstock.com:** ShadeDesign (b). **166 Shutterstock.com:** Beliavskii Igor (Background). **171 Dreamstime.com:** Eugenesergeev (b). **176-177 Shutterstock.com:** Eky Studio (Background). **178-179 Adobe Stock:** Pro500 (Background). **Shutterstock.com:** CESM I Studio

Cover images: *Front:* **Dreamstime.com:** Tartilastock tc/ (X3), Ttretjak c/ (Background); **Shutterstock.com:** Sergey Peterman cr, vi73 br; *Back:* **Dreamstime.com:** Eugenesergeev tr; **Shutterstock.com:** Vonzur bl; *Spine:* **Dreamstime.com:** Martijn De Vries

Endpaper images: Dreamstime.com: Oksana Pasishnychenk

DK LONDON
Editor Millie Acers
Art Editor Anna Formanek
Senior Acquisitions Editor Pete Jorgensen
Managing Art Editor Jo Connor
Production Editor Siu Yin Chan
Production Controller Louise Daly
Managing Director Mark Searle

Written by Adrian Dingle
Designer Neal Cobourne

DK would like to thank Bharti Bedi for copyediting, Caroline Curtis and Elly Dowsett for proofreading and Helen Peters for indexing.

First American Edition, 2025
Published in the United States by DK Publishing,
a division of Penguin Random House LLC
1745 Broadway, 20th Floor, New York, NY 10019

25 26 27 28 10 9 8 7 6 5 4 3 2 1
001–345264–Sep/2025

Published in Great Britain by Dorling Kindersley Limited

ISBN 978-0-5939-6162-9

Printed and bound in China

www.dk.com

This book was made with Forest Stewardship Council™ certified paper—one small step in DK's commitment to a sustainable future.
Learn more at www.dk.com/uk/information/sustainability